Unlocking the Sialic Acid Code: From Structure to Function in Health and Disease

Sanobar

TABLE OF CONTENTS

CHAPTER I

INTRODUCTION

1.1 Sialic Acids (Sias)

The cell surface is covered with different kinds of glycoconjugates, including glycoprotein, proteoglycans, and glycolipids, as observed under electromicroscopy in 1967 [1, 2] (Figure 1. 1) [3]. Sialic acids (Sias) is one of the special monosaccharides commonly located at the terminal of glycoconjugates, which contain nine carbon; based on the difference of substrate linked to the carbon 4-,5-,7-,8- and 9- position, there are more than 50 Sias species is found recently[1, 4-6]. The N-glycolylneuraminic acid (Neu5Gc) and its precursor N-acetylneuraminic acid (Neu5Ac) are the two predominant types of Sias in mammalian cells. Neu5Gc is found in non-human mammalian species due to the mutation of cytidine monophosphate-N-acetylneuraminic acid hydroxylase (CMAH) gene [1, 7], but the intake from dietary such as red meat [8].

Sias could attach galactose (Gal) through α-2,3 or α-2,6 linkage, and N-acetylgalactosamine (GalNAc) or N-acetylglucosamine (GlcNAc) through α-2,6 linkage; or another Sia through α-2,8-linkage [9]. Specific linkages and modification of Sias are involved in the tissue-specific and developmentally regulated manner. Sias is involved in many pathological and physiological processes, including cell adhesion, migration, inflammation, tumor, and tumor progression [10, 11].

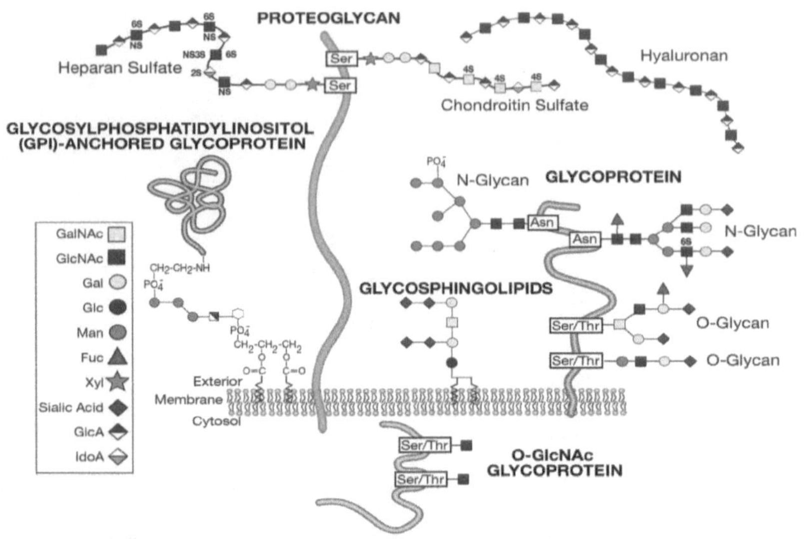

Figure 1. 1 Cell surface glycoconjugates and glycoproteins [3].

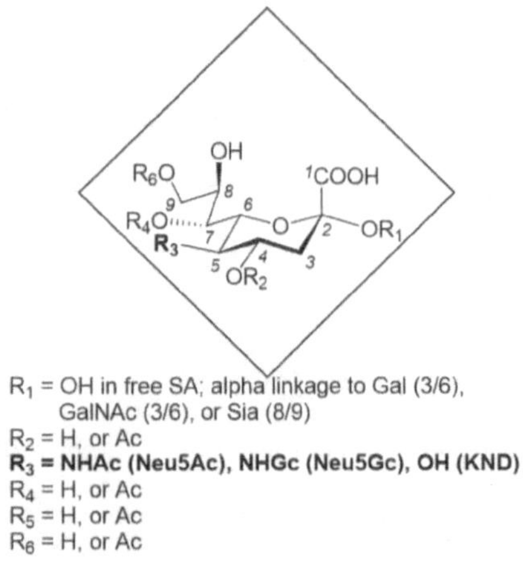

R$_1$ = OH in free SA; alpha linkage to Gal (3/6),
     GalNAc (3/6), or Sia (8/9)
R$_2$ = H, or Ac
R$_3$ = NHAc (Neu5Ac), NHGc (Neu5Gc), OH (KND)
R$_4$ = H, or Ac
R$_5$ = H, or Ac
R$_6$ = H, or Ac

Figure 1. 2 Structures of sialic acids and their linkages in glycans.

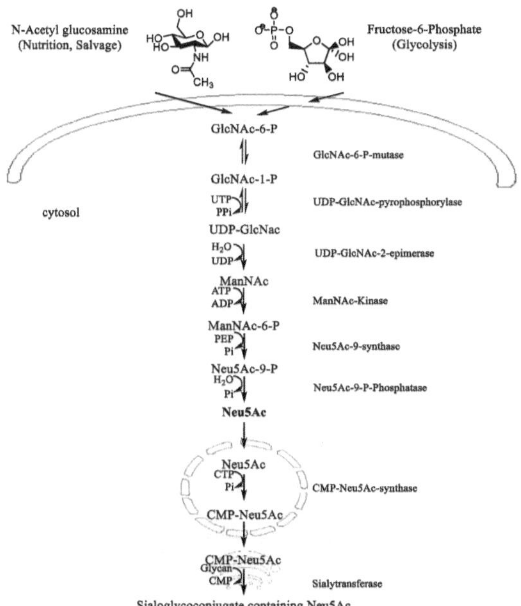

Figure 1. 3 Sias biosynthesis pathway modified from S.J. Moons (2019) [12].

Table 1. 1 Family of sialyltransferases (STs) for specific sialylation linkages modified from J.C. Paulson, C. Rademacher (2009) [13].

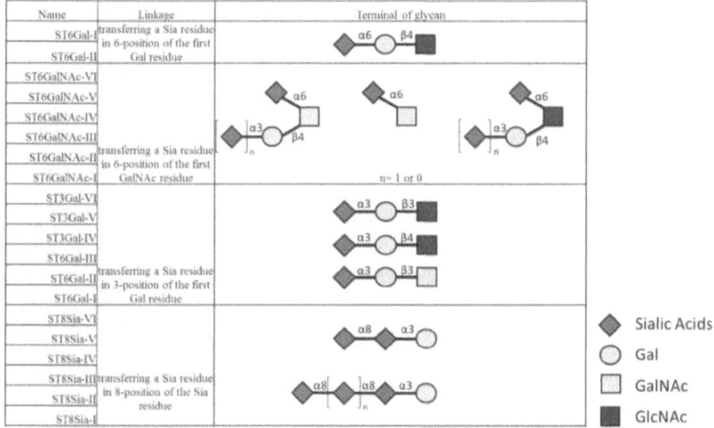

1.1.1 *Sialic acid biosynthesis pathway*

GlcNAc uptake from nutrition or fructose-6-phosphate from glycolysis is converted to GlcNAc-6-P into the cytosol, then the GlcNAc-P-mutase and UDP-GlcNAc-Pyrophosphosphorylase convert GlcNAc-6-P to UDP-GlcAc. The UDP-N-acetylglucosamine-2-epimerase/ N-acetylmannosamine kinase (UDP-GlcNAc 2-epimerase/ ManNAc-kinase, GNE) catalyzes the UDP-GlcAc to form ManNAc-6-P, which has been characterized [14-17]. The ManNAc-6-P is converted to Neu5Ac by Neu5Ac-9-P-synthase (NANS) and Neu5Ac-9-P-phosphatase (NANP), and Neu5Ac is translocated into the nucleus. The Neu5Ac in the nucleus was catalysis by CMP-Neu5Ac-synthase (CMAS) to form CMP-Neu5Ac, and CMP-Neu5Ac is translocated into Golgi. The sialyltransferase is used to add Neu5Ac in Golgi onto the glycoconjugates (Figure 1. 3) [12].

1.1.2 *Role of sialic acids*

Based on the particular location, negative charge, hydrophilicity, and the diversity of linkage on the glycoconjugate, Sias are playing essential roles in the physiological and pathological process, which including immunological process and tumor progression [7, 10, 11, 18]. In several biological systems, Sias contribute to many biophysical features, such as the negative charge of Sias given the charge repulsion to prevent the unwanted interaction of cells in the blood circulation [10]; the density of Sias play a vital role in maintaining the normal filtering function in the kidney [19, 20]; the polysialic acid chains could affect the plasticity of neuron cells [21-24]; Sias is also highly involved in the vascular endothelium [25]. Sias can be used to determine the half-life of glycoproteins in circulation, especially under the inflammation condition [26, 27]; the loss of Sias of glycoprotein will explore another monosaccharide such as galactose and be recognized by the receptors in the liver or other organs, which cause the clearance of glycoprotein[27]. Sias also play a critical role in many pathological processes; Sias measured in the body fluids were used for disease risk prediction. The measurement of Sias in the body fluid are used to predict the disease risk; for example, the total Sias in serum is used for various disease predictions, as many papers suggested [28-31]. There is literature that reported decreased Sias levels on the lipoprotein related to cardiovascular disease risk, Sias levels on the low-density lipoprotein (LDLs) were involved in the uptake of lipid by endothelium and related to the development of atherosclerosis [32, 33]. Sias is highly involved in the innate immune response by interacting with the Siglec (Sialic acid-binding immunoglobulin-type lectins) [34]. Many *in vitro* studies that report a decrease in the Sias level, which affects the immune response [35, 36]. For example, α2-6 linked Sias are highly involved in the development of B cells [37], while the B cell surface molecule CD22/Siglec-2 affects the B cell response to antigen stimulation [38, 39]. The sialoadhesin (Siglec 1) on the macrophage was used to recognize the Sias on the pathogen [40].

1.2 Sialylation and Desialylation

Sialylation and desialylation are two processes of adding or removing Sias from glycoconjugates, respectively. Sias was modified on the surface glycan by a family of enzymes named Sialytransferase (ST) [41]. There are 20 different STs (Table 1. 1) involved in the human glycoconjugates biosynthesis [6]. The sialidase was used to hydrolytic removing Sias from glycoconjugates. Both STs and sialidase play a role in regulating human physiological and pathological pathways[42-45].

1.2.1 *Sialyation*

There are three methods for Sias to get into cells. First, Sias can be taken up by the cells directly from an extracellular fluid; second, Sias can be converted from UDP-GlcNAc in the cytosol; and third, Sias can be recycled from glycoconjugates [46] (Figure 1. 4). Sias are normally located at the terminal of the glycoconjugates via different glycosidic linkages such as α2,3-, α2,6- and α2,8-linkage [3]. The Sias biosynthesis, activation, and transfer to glycoconjugates involve the action in certain enzymes [47]. In the biosynthesis pathway, a GlcNAc from diet and fructose-6-phosphate from the extracellular fluid will be converted to UDP-GlcNAc in the cytosol, UDP-GlcNAc is then converted to ManNAc-6-P by GNE [14-17]. Next, ManNAc-6-P is catalyzed by NANS and NANP to form Neu5Ac in the cytosol[48, 49] and translocated into the nucleus. Sias in the nucleus is activated by CMAS and from the CMP-Neu5Ac [50]. After activation, CMP-Neu5Ac is transported into Golgi and catalyzed by a family sialyltransferase (STs) to form α-2,3, α-2,6, or α-2,8 linked onto glycoconjugates [49, 51, 52]. Sias derivatives such as Neu5Gc and KDN can also be active in CMP-Sias based on the substrate tolerance of CMAS [53, 54], widely involved in the sialo-engineering [55, 56]. STs are family linkage-specific enzymes that are directly involved in the sialylation of glycoconjugates. The ST3Gal1-6 are used to catalyzed C2 of Neu5Ac to the C3 on galactose (Gal); ST6Gal1,2 linked C2 of Neu5Ac to the C6 of Gal; C2 of Neu5Ac is linked to C6 of GalNAc by ST6GalNAc; ST8Sia1-6 is used to link C2 Sias to another Sias C8 position [3] (Table 1. 1) [13].

1.2.2 *Desialylation*

On the other hand, Sias on the glycoconjugates could be removed by sialidase (neuraminidases, Neu). In mammalian cells, there are four sialidases involved in this process, which are Neu1, Neu2, Neu3, and Neu4 [57, 58]. Neu1 is the lysosomal sialidase, which is used to degrade the sialoglycoconjugates [59-61]; Neu2 is the cytosolic sialidase which is highly involved in removing gangliosides in the cytosol [62, 63]; the membrane-associated Sialidase Neu3 is the enzyme located at the membrane and specific for gangliosides [64]; Neu4 is found on the outer membrane of mitochondria, which work on a variety of glycoconjugates including glycoproteins, gangliosides, and oligosaccharides [65] (Table 1. 2) [66]. Sias is released in the cytosol or the extracellular fluid and pumps back to the cytosol, and the released Sias is used in another cycle of Sialoglycoconjugate formation [46]. Sias is also widely involved in viruses and microorganisms, but not in plants, insects, or yeast [67, 68].

1.2.2.1 Neu1

Neu1 is a lysosomal Sialidase used to remove Sias located at the terminal of sialo-glycoconjugates [60]. The Neu1 is the most widely-studied sialidase, as Neu1 deficiency is related to many diseases such as sialidosis and galactosialidosis [59]. Sialidosis caused by the Neu1 deficiency commonly occurs before birth, which causes the stillborn of patients or death after birth [69]. The galactosialidosis could also be caused by Neu1 deficiency coupled with protective protein/cathepsin A (PPCA) deficiency [44]. Neu1 not only works in the lysosomal under the acidic conditions but also works as a membrane boundary protein on the plasma membrane, which is controlled by the phosphorylation of the C terminal of the protein [69]. At the plasma membrane, Neu1 plays essential roles in signal transduction, elastogenesis [70-72], modulation of insulin receptor signaling, and regulation of many receptors [73-80]. In the lysosome, Neu1 works as a multienzyme complex by coupling with PPCA and β-galactosidase [81]. In this complex, the PPCA works as the chaperone for Neu1, preventing Neu1 premature oligomerization [82]. The PPCA/Neu1 complex is also required, localizing Neu1 to the lysosome for the activities as a lysosomal enzyme [83]. However, Neu1 translocation to the plasma membrane is coupling with or without PPCA still unknown [60]. Neu1 is proposed to be involved in many essential functions in cells, such as degradation in the lysosome, exocytosis, immune function, elastic fiber assembly, and carcinogenesis.

Neu1 was found to negatively regulate the lysosomal exocytosis process, which is a cellular process fusing the lysosome to the plasma membrane [84]. The exocytosis of lysosomes was found in many physiological and pathological processes such as the replenishment of the plasma membrane in macrophages, damaged plasma membrane repair, and removal of the pathogenic bacteria or infected virus from the infected cells [85-89]. Neu1 knock-out mice were used to mimic the human condition, which shows increased lysosomal proteases, glycosidases secretion, and the presence of lysosome-associated membrane protein 1 (LAMP-1) [90]. LAMP-1 is a heavily sialylated protein that is the target substrate of Neu1 [91]. Decreased the expression of LAMP-1 in the cells limits the lysosomal exocytosis [84]. The down-regulation of Neu1 caused the oversialylated LAMP-1 and enhanced lysosomal exocytosis of soluble hydrolases and exosomes [92]. As the expression of Neu1 is related to lysosomal exocytosis, the Neu1 expression also could be developed as the diagnostic biomarker and serve as the potential therapeutic target for aggressive and untreatable human pleomorphic sarcomas [92].

Neu1 plays an essential role in the immune system. Sias exist at the terminal of glycoconjugates and are involved in many physiological processes such as immune responses, and the balance between sialidases and sialyltransferases commonly controls the sialylation of the sialo-glycoconjugates [44, 93]. Sialidase treatment increases the capacity of resting B cells to stimulate the proliferation of allogeneic and antigen-specific, syngeneic T cells so that the cell surface Sias were identified as the potential modulator of immune cell interaction [94]. The resting B cells are poor antigen-presenting cells (APC) with less ligand present on the B cells that are required for the T cell activation [95-97]. The cell surface Sias are involved in the B cell activations. The sialidase treatment of the resting B cells significantly enhances the presentation of antigen to allogeneic T cells by the loss of 2,6- linked Sias on the surface of B cells [94]. Moreover, there are more studies showing the Neu1 on the surface of activated T cells are used in the cell surface desialylation and cause the production of cytokines such as inerleukin 4 (IL-4) and interferon γ (IFN-γ) [36, 98]. Neu1 is also required to interact with T cells and antigen-presenting cells and convert the group-specific component protein into a factor necessary for the inflammation-primed activation of macrophages [99, 100].

Many studies determined the Neu1 activity was increased 20-30 folds during the monocytes differentiation to macrophage in blood circulation; on the other side, the Neu2, Neu3, and Neu4 activity remains the same or even decreased [35, 101, 102]. In the differentiation process, the activated Neu1 complex in the lysosome was found to be sorted to the plasma membrane through the LAMP-2 negative vesicles [35]. There is another research that found macrophage and immature dendritic cells from Neu1-deficient mice presenting more Sias on cell surfaces and enhanced the ability of phagocytosis [103]. The cell surface sialylation was significantly decreased, and phagocytosis function was restored when the cells were treated with purified mouse Neu1. Neu1 caused desialylation affecting the transduction of signals from Fc receptors of Immunoglobulin G [103]. Neu1 also regulates immune receptors such as Toll-like receptors (TLR), which plays an essential role in inflammation [76, 104]. The ligand, such as lipopolysaccharides (LPS), binds to the TLR receptor, which induces Neu1 activity and promotes intercellular signaling and cytokine release [76, 104, 105]. TLRs are sialoglycoproteins, and Neu1 removed the Sias from the TLRs could active the inflammation function of cells [106].

Neu1 is required for the normal assembling of elastic fibers. The assembling of elastic fibers in tissues is under the repeated cycle of extension and recoil [107]. The elastic lamellae in the aorta of Neu1-null mice were found to be thinner and separated by hypertrophic smooth muscle cells, with the elastin level being significantly reduced in the aorta of Neu1-null mice [107]. Neu1-null mice have been generated as an accurate sialidosis model, which has Neu-1 deficiency affecting elastin metabolism. The elastin-binding protein (EBP) is an enzymatically inactive splice variant of lysosomal β-Gal [108]. Neu1, PPCA, and β-GAL are formed the active complex in the lysosome [81] and cause the reassembling on the cell membrane. The cell surface Neu1 activates the EBP tropoelastin complex by removing the terminal Sias from glycoconjugates [109].

Cancer cells and tissues show deceased Neu1 expression compared to adjacent non-cancerous counterparts. An inverse relationship was found between Neu1 expression and metastatic ability in various cell types. In rat 3Y1 fibroblasts, a Neu1 decrease was associated with higher metastatic potential [110]. In mouse adenocarcinoma colon 26 cells, the decreased expression of Neu1 was found in highly metastatic NL17 and NL22 cells [111]. In the B16-BL6 murine melanoma cells, more expression of Neu1 was found in a highly invasive and metastatic line [112]. In a previous study, an inverse relationship was found between sialidase Neu1 expression and metastatic potential of murine cancer cells

[75]. The colon cancer HT-29 cells were transiently or stably transfected with the Neu1 gene. *In vivo*, liver metastasis was significantly reduced when the mice were injected with the Neu1-overexpressing cells. The major molecule associated with decreased sialylation is integrin β4, which was accompanied by decreased phosphorylation of the integrin followed by attenuation of focal adhesion kinase and extracellular signal-regulated kinase 1/2 (Erk1/2) pathway. Neu1 also caused the downregulation of matrix metalloproteinase-7 that is associated with cancer metastasis [75]. However, Neu1 does not always act as a cancer suppressor as a higher Neu1 expression level was found in ovarian cancer tissues than adjacent non-cancerous tissues. Human ovarian cancer cells were prevented with the inhibition of Neu1 by siRNA [113]. Neu1 is activated by an epidermal growth factor receptor (EGFR)-induced G protein-coupled receptor signaling process and metalloproteinase-9 activation. The active Neu1 in the complex hydrolyzes α-2,3 Sia residues on the receptors, which removes the steric hindrance of receptor association and allow subsequent dimerization, activation, and signaling [114-116]. The oseltamivir phosphate is a sialidase inhibitor which inhibits Neu1 activity, which extramly decrease the cancer cell survival rates through an intrinsic signaling platform in human pancreatic cancer with acquired chemoresistance [115, 116].

1.2.2.2 Neu2

Neu2 is a functional sialidase located in the cell cytosol that works at an optimum pH of about 6.0-6.5. Neu2 cDNA first was isolated from the rat skeletal muscle [117], and homologs were isolated from CHO [118], mouse brain [119, 120] and mouse thymus [121], and human skeletal muscle [63] showing high identity to the rat gene. The first human Neu2 three-dimensional structure was determined by X-ray crystallography [62], being a canonical six-blade beta-propeller with an active site in the shallow crevice. The kinetic characterization was tested with purified human recombinant Neu2 (hsNEU2) from the overexpression homologue in *E. Coli* [122]. The highest catalytic activity and affinity was found for α-2,3 linked sialoglycoconjugates, but while the α-2,8 linkage in colominic acid and the α-2,6 linkage carried by α-2,6 sialyllactose were resistant to hydrolysis [122]. hsNEU2 works on gangliosides, such as GD1a, GM1, and GM2, at nanomolar concentrations. When the HsNEU2 works with gangliosides substrate in the buffer with the detergent (Triton X-100), the V(max) will be increased.

Neu2 is involved in the differentiation of skeletal myoblasts *in vitro*, as an increased Neu2 gene transcription is noted in rat L6 myoblasts [123] and murine C2C12 [124] myoblasts. However, Neu2 showed downregulation in the SJL mouse, the human dysferlinopathy model [125]. The downregulation of Neu2 may contribute to impaired muscle regeneration [125]. In PC12 cells, Neu2 is involved in neuronal differentiation by activating the nerve-growth factor-induced transcription [126]. Neu2 also drives insulin-like growth factor 1- induced hypertrophy of myoblasts [127]. Neu2 enzymatic activity is linked to proper muscle differentiation and growth, as Neu2 is activated by AKT or Igf-1in myoblasts [128]. Neu2 in human cancer cells was investigated and exhibited significant effects on cell growth and apoptosis [129]. The transfection of the CHO Neu2 gene into the A431 human epidermoid carcinoma cell line caused reduced GM3 levels and enhanced cell growth and tyrosine autophosphorylation of EGF receptor at lower concentrations [130]. Adding the human Neu2 gene into K562 leukemic cells caused a remarkable reduction in anti-apoptotic factor Bcl-XL and Bcl-2, resulting in increased sensitivity to apoptotic stimuli [131]. The overexpression of Neu2 in cells reduced gene expression and activity of Bcr-Abl, as well as Bcr-Abl-dependent Src and Lyn kinase activity through desialylation of some cytosolic glycoproteins. However, due to the fact that the expression level of Neu2 in human tissues and cells are extremely lower than other sialidases [132], the Neu2 function is still unknown.

1.2.2.3 Neu3

Neu3 is a plasma membrane-associated sialidase, which was first cloned from a bovine brain library, later cloned from a human skeletal muscle cDNA library [133] and human genome database [134]. The primary sequences are highly identical within the human gene, mouse gene [120], rat gene [135], and bovine gene [133]. Neu3 from bovine and humans specifically hydrolyze gangliosides, except for GM1 and GM2. The murine Neu3 works on oligosaccharides, 4MU-NANA, and glycoproteins.. The human Neu3 was also detected in other intra-membranous components other than the plasma membrane [129]. For this reason, gangliosides such as GM3, GD3, GD1a, and GD1b are suitable substrates for the *in vitro* study for Neu3 [129, 136]. In response to EGF, Neu3 redistributed rapidly to membrane ruffles coupled with Rac-1 in A431 cells [137]. Neu3 is involved in neurite formation [138] and the regulation and regeneration of rats hippocampus neurons [139, 140]. Neu3 was also found in rafts of neuroblastoma cells [141] and the caveolae of HeLa cells coupled with caveolin-1 [142]. Neu3 controls the transmembrane signaling by reacting with signaling molecules such as caveolin-1, Rac-1, interin β4, Grb-2, and EGFR to catalyze the gangliosides [129].

Based on the particular location and strict substrate performance of Neu3, Neu3 plays an essential role in regulating cell surface function. Neu3 was also found in rafts of neuroblastoma cells [141], and the caveolae of HeLa cells coupled with caveolin-1 through the caveolin-binding region [142]. In caveolae, Neu3 was used to modify the GM1 level at the cell surfaces used to control platelet-derived growth factor (PDGF) -induced Src mitogenic signaling and DNA synthesis [143]. The phosphoprotein associated with glycosphingolipid-enriched microdomains (PAG) is a ubiquitously expressed member of the transmembrane adaptor protein family, which is well known to downregulate SRC family protein tyrosine Kinases (SFK) by binding to the Csk [143]. The ganglioside GM1 level at the cell surface was increased by PAG-N not binding to Csk, which is required for the activity of Neu3. Neu-3 controls PAG-N – induced GM1 antimitogenic effects; the results were confirmed with the DANA sialidase inhibitor, or the Neu3 siRNA inhibition reversed the PAG-N mitogenic inhibition [143]. Neu3 is also found in tetraspanin-enriched microdomains (TREMs) incorporated in various transmembrane receptors, modulating their activities [144]. In many studies, gangliosides are related to tetraspanins and are involved in regulating signaling [144-147]. Overexpression of Neu3 in mammary epithelial cells affected the stability of CD82-containing TERMs, which show the reduced association with CD151 and enhanced association with EGFR [144]. Neu3 was identified as a physiological regulator of the integrity of these microdomains by regulating the gangliosides.

Neu3 also plays an essential role in the proliferation and differentiation of neuronal cells [148]. The mouse ganglioside sialidase cDNA was cloned and expressed in the Neuro2a cells to understand the role and regulation mechanisms of sialidase in the neuronal systems. The presentation of Neu3 in the Neuro2a cells showed an acceleration of neurite arborization following 5-bromodeoxyuridine-induced Neuro2a cell differentiation [120]. More studies show Neu3 is involved in axonal growth and regeneration by regulating the

GM1 level [139, 149]. The Myelin-associated glycoprotein (MAG) was shown to bind to the nerve cell surface and inhibit nerve regeneration; the nerve cell surface gangliosides GD1a and GT1b serving as MAG ligands, and were involved in the inhibition of nerve regeneration [150]. Neu3 was used to remove Sias from polysialylated gangliosides GD1a and GT1b to form monosialylated ganglioside GM1, which was shown to inhibit MAG binding to the nerve cell surface and enhance axon regeneration and functional recovery in the spinal cord injury rats [149, 151]. Neu3 is asymmetrically located at the tip of one neurite of the unpolarized rat neuron, which could cause the actin instability. The destabilization of actin triggers the polarization of neurons [140]. Suppressing Neu3 activity blocks axonal generation and enhances single axon formation [140]. Enhancing TrkA activity could cause Neu3 induced axon specification. This could trigger phosphatidylinositol-3-kinase (PI3K)- and Rac1-dependent inhibition of RhoA signaling and consequent actin depolymerization in one neurite [140].

Neu3 also plays a crucial role in skeletal muscle differentiation by modulation of the ganglioside content GM3 [152-154]. The down-regulated Neu3 in the murine C2C12 myoblasts inhibited the capability to differentiate by increasing the GM3 level above a critical point, leading to the EGFR inhibition and enhancing the responsiveness of myoblasts to the apoptotic stimuli [152]. The overexpression of Neu3 affected the EGF signaling pathway that promoting cell proliferation and delaying the beginning of the differentiation of myoblast [153, 154]. Because the up-regulation of Neu3 causing the suppression of cell apoptosis in human cancers by removing Sias in the gangliosides and increased activation of the mitogenic receptor, Neu3 was proposed as a cancer marker [155-157].

1.2.2.4 Neu4

Neu4 is found in the lysosomal lumen, mitochondria, and endoplasmic reticulum (ER) [103, 158, 159]. Neu4 have two iso-forms, in human, Neu4L, and Neu4S, which only have a 12 amino acid difference on N-terminal [158]. The isoforms Neu4 are also expressed differentially in a tissue-specific manner; two isoforms were found in the brain, muscle, and kidney, while the Neu4S isoform is found in the liver and colon [158]. In mice, Neu4 exists as Neu4a and Neu4b, in which Neu4a has an additional 23 amino acids at the N-terminal compare to Neu4b [160]. Neu4 works with diverse sialo substrates like Neu2. However, it is the only sialidase that hydrolyzes mucins. Neu4 plays a regulatory role in the ganglioside catabolism in the mice nervous tissue, in which increased GD1a and decreased GM1 were found in the Neu4 deficient mice [161]. The expression of Neu4 is predominant in the mouse brain, which has a significantly higher level than Neu3 [162]. Human Neu4L is localized to mitochondria and specifically expressed in the brain, affecting the level of GD3 in neuroblastoma cells [163]. The GD3 acts as the apoptosis-inducing ganglioside [164]. Unlike Neu3, Neu4 appears to down-regulate neurite formation in the Neuro2a cells during the differentiation [129, 160]. PolySia is involved in the control of synaptic plasticity, neuronal differentiation, and cell migration [24, 165]. Mouse Neu4 was found to hydrolyze polySia on neural cell adhesion molecules (NCAMs) in the hippocampus [162]. However, the mouse Neu4 gene is expressed dominantly in the brain but only in extremely low amounts in other tissues [166]. The mRNA level and expression level of Neu4 were found to decrease in human colon cancer compared to non-cancerous tissue [167]. The transfection of Neu4 into DLD-1 and HT-15 colon adenocarcinoma cells resulted in the acceleration of apoptosis and inhibited invasiveness and cellular motility [168].

Unlike Neu1, Neu4 targets gangliosides GM2 in the lysosome lumen,. Neu4 might be used as the new therapies development for sialidosis and galactosialidosis, which have decreased the expression of Neu4 [161, 169]. A marked vacuolization and lysosomal storage in lung and spleen cells were observed in the Neu4 knock-out mice model [170]. In the knock-out Neu4 mice model, an increased level of gangliosides, ceramide, cholesterol, and fatty acid was observed in the brain, liver, lungs, and spleen [161]. The HexA deficiency could be corrected by the glycol lipid GA2 by correcting the impaired metabolism of GM2 [161].

Neu4L located in the mitochondria plays an essential function in the brain. Unlike Neu3, Neu4 negatively regulates neurite formation in the Neuro2a cells during the differentiation by removing Sias on polySia [129, 160, 170]. As a result, overexpression of Neu4 will result in suppression of neurite formation, and while a knockdown of Neu4 will cause the acceleration of neurite formation [160].

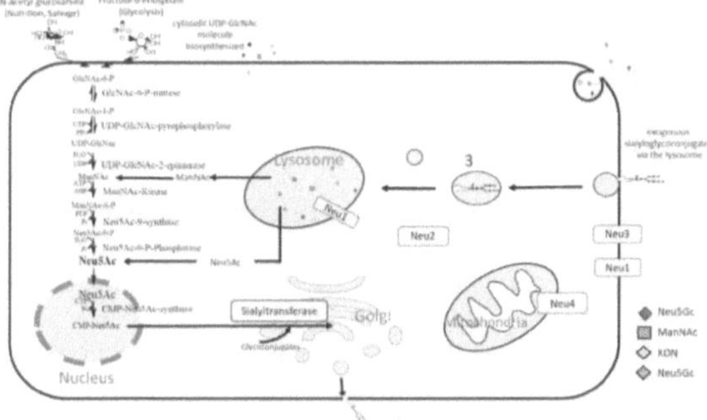

Figure 1. 4 Biosynthesis pathway of Sias and sialidases and their subcellular locations modified from S.J. Moons (2019) [12].

Table 1. 2 General properties of four mammalian sialidases [66]

	NEU1	NEU2	NEU3	NEU4
Major subcellular localization	Lysosomes	Cytosol	Plasma membranes	Lysosomes[a] Mitochondria[a] and ER
Good substrates	Oligosaccharides Glycopeptides	Oligosaccharides Glycoproteins Gangliosides	Gangliosides	Oligosaccharides Glycoproteins Gangliosides
Optimal pH	4.4–4.6	6.0–6.5	4.5–4.7	4.5–4.7
Total amino acids				
Human	415	380	428	496 (484)[b]
Mouse	409	379	418	501 (478)[b]
Chromosomal location				
Human	6p 21.3	2q 37	11q 13.5	2q 37.3
Mouse	17	1	7	10
Proposed major functions	Degradation in lysosomes Exocytosis Immune function Phagocytosis Elastic fiber assembly	Myoblast differentiation Neural differentiation	Neuronal differentiation Apoptosis Adhesion	Neuronal differentiation Apoptosis Adhesion

1.3 Sialic Acids-Related Diseases

Vertebrate cells are coated with a dense and complex array of sugar chains attached to proteins and lipids. Most soluble secreted proteins also contain post-transcription modifications, such as glycosylation [3-6]. Sias is a diverse family of sugar units that contains nine carbon backbones and are located at the terminal position of glycans. Because of this specific location and ubiquitous distribution, Sias was involved in many physiological and pathological processes [10, 11]. The altered glycosylation on the cancer cells was found in various cancer types and stages. Elevated levels of Sias (hypersialylation) are found on cancer cells compared to the normal cells [171, 172]. Neu5Gc containing glycans may promote inflammation and cancer progression [18]. Serum Sias have been used to predict coronary heart disease, stroke mortality and reflect the existence or progression of an atherosclerotic process [173]. The reduced level of Sias on platelets, erythrocytes, and lipoproteins may play an essential role in the pathogenesis of atherosclerosis [174-176].

1.3.3 *Hypersialylation in cancer*

Sias are expressed by normal healthy cells. However, an abnormally high level of Sias has been observed on tumor cells [171, 177]. Hypersialylation plays an essential role in immune evasion, metastasis, and intracellular interaction [171, 177]. The hypersialylation of membrane-bound glycoconjugates inhibits the Sias-binding receptors such as Siglec receptors and activates the NKG2D ligand, and causing a reduced activity of the Natural Killer (NK) cells [177, 178]. Three different possible mechanisms might elevate levels of Sias; altered levels of STs and the level of neuraminidases or altered levels CMP-Sias for STs.

Studies have been shown that malignant disease states are associated with at least 9 of 20 STs. ST6GalI is expressed normally in all homeostatic tissue, while been up-regulated in various cancer tissues, such as the colon, breast [179-181]. Elevating the level of ST6GalI may be associated with a poor prognosis and the Ras oncogene. Increased ST6GalI expression level may cause cells resistant to tumor necrosis factor (TNF)-induced apoptosis, proposed to be mediated through the addition of Sias on TNFR1 death receptor [182], as well as affecting the tumorigenesis through blocking binding sides for Galectin to inhibit the apoptosis [183, 184] and modulation of immune cells [185]. ST3GalI-VI was used to add Sias on the Gal residue in an α2,3- linkage. ST3GalI added Sias on the Core-1 O-glycan, which could help mask the tumor from surrounding immune cells [186]. The expression level of ST3Gal3 was increased in a variety of cancers, and it was linked to cancer progression [187-189]. ST3Gal4 and ST3Gal6 are involved in upregulation in many different cancer types, which are linked to an increased metastatic and invasive disease state as the sialoglycoconjugates bind selectin expressed on inflamed vascular endothelial cells [190-195]. The ST3Gal4 and ST3Gal6 modified glycans also target regulatory T-cells to suppress localized tissue inflammation [196]. ST6GalNAcI-VI transfers Sias to the GalNac residue on glycan through an α2,6- linkage. ST6GalNacI adds Sias on the Tn antigen; the overexpressed Sialyl-Tn-antigen (STn) being widely recognized in a variety of cancer types [197, 198]. The overexpressed STn has been shown to induce less Th1-induced pro-inflammatory cytokines and increase the invasiveness of cancer cells in a MUC1-dependent manner [199-201]. However, downregulated ST6GalNAcII is elevated the malignancy by switching the type of sialylated glycans presenting [202]. ST8SiaI-VI are sialyltransferase used to linked Sias on Sias residues through α2,8- linkage. The elevated ST8SiaI and ST8SiaIV levels are involved in cancer cell etiology and pathophysiology in many cell types [203]. The overexpressed ST8SiaII and IV are associated with elevated astrocytoma tumors because the highly expressed polySia is linked to the tumor invasiveness [204]. The increased ST8SiaII level causes increased phosphorylation levels of ERK1/2 (Extracellular signal-regulated kinases), MMP-9 (matrix metalloproteinases), and FGFR1(Fibroblast Growth Factor Receptor 1), leading to invasion and migration of cells *in vitro* [205]. Overexpressed ST8SiaIV level is involved in increased cell survival by regulating the PI3K/Akt pathway [206].

Four mammalian neuraminidases (Neu1-4) are involved in the various cancer types. The Neu1 has been reported linked to cancer metastasis, as it is involved in facilitating the epithelial to mesenchymal transition [115]. Overexpressed Neu2 is linked to the increased cell survival in prostate cancer [207]. The overexpression of Neu2 shows increase tumor cell apoptosis in a leukemia cell line, which is controversial [131]. Overexpressed Neu3 has been observed in various cancers such as prostate cancer [208], melanoma [209, 210], and ovarian cancer [211]. Although overexpressed Neu3 is expected to cause hyposialylation, the overexpressed gangliosides in various cancer types allow protection from immune cells, causing cancer cell invasiveness and metastases [212-214]. A down-regulated Neu4 leads to increased tumor metastasis [168].

The CMP-Sias and underlying glycoconjugates are other mechanisms affecting the level of cell surface Sia. Due to the metabolic flux of Sias, the cancer cells become more aggressively malignant by modulating the sialylation of proteins implicated in metastatic transformation [215]. One of the metabolic pathways that changes CMP-Sias is varying the level of UDP-GlcNAc [215], which can alter N-glycosylation patterns of sialylation [216]. The increased level of UDP-GlcNAc also leads to an increase in the transcriptional level of ST, which elevates the sialoform proteins, including proteins involved in metastasis in cancer [217]. Recently, the intracellular level of Sias in the cancer cells was shown to be dependent on the stage of cancer [218].

1.3.4 *Serum sialic acids in cardiovascular disease*

Serum Sias has been shown to be a strong predictor for cardiovascular mortality and may also reflect the atherosclerotic process [173]. Many studies have shown an elevated level of serum Sias for cardiovascular diseases, such as carotid atherosclerosis, coronary heart disease (CHD), stroke, and peripheral vascular diseases [174, 219-227]. Increased level of Sias correlated with cytokines such as tumor necrosis factor (TNFα) and IL-6 that regulate C-reactive protein (CRP) and the synthesis of other acute-phage proteins [228]. Increased CRP was shown to indicate a higher risk for coronary heart disease, sudden death, and peripheral arterial disease [229]. The mean serum level of glycoproteins was found significantly higher in carotid atherosclerosis [230]. LDL plays an essential role in the pathogenesis of CHD. Sias on LDL was found to be significantly lower in CHD patients [231-233]. The desialylated LDL was atherogenic characterized by small size, higher density, and increased electronegative charge compared to the native LDL [234]. The desialylated LDL was found in the arterial wall atherosclerotic lesion [235]. Also, removing Sias from the endothelial cell surface was found to increase the rate of receptor-mediated endocytosis of LDL by altering the Sias level of the LDL receptors [236]. Desialylation of LDL increased the rate of metabolic clearance [237], internalized the lipoprotein, and regulated cellular cholesterol metabolism [237]. The LDL Sias content was increased in patients with both coronary stenosis and peripheral arterial atherosclerosis lesions, which might be caused by the peroxidation of LDL due to the presence of EDTA in the buffer [238].

1.3.5 *Neu5Gc*

Neu5Gc is one of the major forms of Sias in most mammals; however, it is not biosynthesized in humans who do not express missing CMAH [1, 9]. The incorporated Neu5Gc found in human glycans comes from dietary sources. The intracellular Neu5Gc is hydrolyzed from exogenous glycoconjugates or free Neu5Gc and converted to CMP-Neu5Gc and used as the substrate for STs [239]. Xenoautoantibodies have been reported to be linked to inflammation and tumorigenesis. In which anti-Neu5Gc antibodies are produced by humans responding to Neu5Gc (241). The xenoautoantibodies could drive inflammation and the pro-inflammatory cytokine IL-6 production [240]. The $Cmah^{-/-}$ (CMP-Neu5Ac hydroxylase knockout) mice with a human-like deficiency of Neu5Gc, the $Cmah^{-/-}$ mice were fed Neu5Gc rich glycoprotein, resulting in an increased rate of liver cancer [240].

1.3.6 *Sialic acids in the immune system*

The high density of terminal Sias on the glycoconjugates contributes negative charge and hydrophilicity to cell surfaces, which could alter biophysical properties [24, 241]. Many studies show desialylation of the immune cell surface, which could affect the behavior of cells and many properties [242]. When the cell surface Sia bind to the major serum protein factor H, which could protect cell surfaces from the alternative complement pathway by downregulating the constant "tick-over" [243-246]. The intrinsic Sias binding to selectins could also modulate leukocyte trafficking [247-251]. Sias binding Siglec could control the immune cell activation, with desialylation shown to enhance the immune cell activation [34]. The host's Sias could be used to recognize and bind to the pathogens such as protozoa, viruses, bacteria, and toxins [252-256]. Microbial mimicry of host Sias allows triggering of the host immune response by inhibiting the complement pathway [257-259].

1.4 Sialic Acids and Macrophages

Macrophages are the bone marrow-derived cells, which played an essential role in the immune system in steady-state conditions or inflammation [260]. The macrophage can be further polarized to classically (M1) or alternatively (M2) form [101]. The M1 macrophage is associated with acute inflammatory responses, and M2 is related to the inflammatory responses and adaptive type I immunity [261-263]. Many receptors were located on the surface of the macrophage. One group receptor is sialic acid-binding immunoglobulin-type lectins (Siglec), which recognize the sialic acid from the same cells or other cells, and even pathogens. Another group receptor is the sialylated receptor, such as TLR. Desialylation of TLR could cause the activation of macrophages.

1.4.1 *Siglec on macrophage*

The macrophage is covered with various receptors for the activity and function of the macrophage, including differentiation and phagocytosis [264]. The most common recognition receptor found on the macrophage is Siglec, with some subfamilies of Siglec being involved in recognizing the Sia on the pathogen [265]. The different Siglecs, preferentially bind to the different linkages on glycoconjugates [266-268]. For example, sialoadhesin preferentially recognizes Neu5Acα2→3 galactose glycans [267]. The desialylation of parasites has been shown to decrease its recognition of the macrophage [269]. There is increased uptake of desialylated LDL by LPS- stimulated macrophage from blood [270].

1.4.2 *Sia level in macrophage*

The macrophage cell surface is covered with a dense layer of glycan, which affects its functions and activities [271]. Sias are found on the terminal of many glycoconjugates, which including glycolipids and glycoproteins [1]. Meanwhile, there are many cell surface receptors that are sialylated, such as the Toll-Like receptors. The Sia level on the macrophage cell surface influences the cell's function, properties, and activities [101, 272]. The sialylation levels were changed during THP-1 monocytes differentiation and THP-1 macrophage polarization, with the change of Sia levels being related to the function of macrophages, such as the inflammatory response[101]. The desialylation of the macrophage was found to enhance inflammatory responses [35, 103, 273-277]. The reduced level of the Fc fragment on Immunoglobulin G could cause an increase in the antigen-specific immune response [274]. The macrophage was desialylated during the virus infection, causing the enhanced phagocytosis function [278]. The cell surface desialylation of the macrophage by viral sialidase could enhance the macrophage's activity for phagocytosis of virus-infected HeLa cells [279]. The desialylated TLR4 caused the release of Siglec E and enhanced and maintained the inflammatory activation of the macrophage [280].

1.5 Methods used for Sias Qualification and Quantification

Sias are monosaccharides generally found as the terminal unit on glycoconjugates that coated cell surface [10]. Sias play an important role in many different biological processes such as cell interaction, immune response, as well as normal cell function [10, 11]. Sias serves as a biomarker for many diseases such as cardiovascular disease [173, 174], diabetes [281], and different cancer [282-285]. Sias levels are increased in CHD, diabetes, and cancer patients compared to healthy individuals [286]. Four assays have been developed for Sias qualification and quantification: colorimetric, fluorometric, enzymatic, and chromatographic/mass spectrometry.

1.5.1 *Colorimetric methods for quantification of Sias*

The colorimetric methods were the first methods developed for Sias quantification in biological samples. These methods normally require the Sias to be released from the sialoglycoconjugates by mild acid hydrolysis, and then followed by a chromophore formation, with the absorbance at a certain wavelength for Sias quantification. Reagents used for chromophore formation are diphenylamine [287] and dimethylaminobenzaldehyde [288]. Mild acids used for hydrolysis include sulfuric acid/acetic acid [289], HCl [290], ethanoic acid/sulfonic acid [281]. However, these methods were found to have low specificity and sensitivity for Sias. Hexoses, pentoses, deoxy sugars, and ketohexoses could also be detected under these assay conditions. Due to the low sensitivity and specificity, most of these methods could only be used to detect the total Sias in biosamples, with the exception of Periodic acid/Methyl-3-benzothiazolinone-2-hydrazone (MBTH) assay. The colorimetric assays utilize orcinaol, resoricinal, triobarbiturate, and MBTH reagent. The orcinaol assay was reported in 1941 by Klenk et al. [291]. The mild acid was used to remove Sias from Sialoglycoconjugates. Sias is boiled at 100 °C in the Bials reagent (40.7 mL concentrated HCl, 0.1 g orcinol, 1 mL 1% FeCl$_3$, and 50 mL water) to form the intermediate ketone [292]. The ketone reacts with orcinol to form the chromogen, which is extracted from the aqueous solution to give the chromophore [293], having a maximum absorbance at 572 nm. The resorcinol method is built on the orcinol method, which was reported in 1957 [294]. Sias is reacted with resorcinol reagent in 10 mL of solution (2 g resorcinol in 100 mL water and adding this to 80 mL of concentrate HCl containing 0.25 mL of 0.1 M copper sulfate solution). Sia and reagent were heated to 110 °C for 15 min to form the chromophore, which has a maximum absorbance at 580 nm [294, 295]. The triobarbiturate method was reported in 1959. The periodic acid was used to oxidize Sias to form the aldehyde in the acidic solution, 0.125 M sulfuric acid as developed by Aminoff [296], or 9 M phosphoric acid by Warren [297]. The aldehyde reacted with thiobarbituric acid under the acidic condition for production of the chromophore with maximum absorbance at 550 nm, extracted with cyclohexanone by Warren [297], or acidified butanol by Aminoff [296]. Until 1979, the MBTH was reported by Massamiri [298]. This method was used to measure the bound Sias without acid hydrolysis. Formaldehyde was formed from the kinked Sias by reacting it with periodic acid in phosphate-buffered saline. The formaldehyde was reacted with zinc sulfate, and 1M NaOH followed by MBTH reagent in dilute HCl and FeCl$_3$ to form the chromophore having maximum absorbance at 625 nm. Since this method does not need to cleave Sias from sialoglycoconjugates, it could be used to detect both conjugate Sias and free Sias.

1.5.2 *Fluorometric methods for the quantification of Sias*

Fluorometric methods for the Sias qualification and quantification have involved the reaction between Sias with an oxidant or acid to form a fluorophore with certain excitation and emission wavelengths. These methods are useful for the Sias measurement but are not able to distinguish different species of Sias. The fluorometric methods give more specificity and sensitivity than the colorimetric methods [286]. The 3,5-diaminobenzoic acid method is the first developed fluorometric method Sias, which was reported in 1964 [289]. Sias is reacted in 0.005M 3,5-diaminobenzoic acid, and 0.125M HCl at 100 °C for 16h. This method has high specificity for Sias, which provides accurate results in quantifying Sias in samples. The thiobarbiturate method was slightly modified for fluorometry in 1976 [299]. Sias released from sialoglycoconjugates by mild acid hydrolysis and then is reacted with thiobarbituric acid to form a fluorophore, which extracted into the acidified butanol to get the fluorophore with ex/em=550/570 nm. This method was 500 more sensitive than the thiobarbiturate colorimetric method, which could be used to detect picomole Sias. The pyridoxamine method was developed in 1976 by Murayama et al., which is a more sensitive method that could detect Sias as low as 0.1 µg [300]. Pyridoxamine dihydrochloride reacted in zinc acetate and pyridine in methanol heated at 70 °C for 45 min, and reacted with Sias and form a ring-opening form Sias to form the fluorophore ex/em=395/470 nm. This method is specific for the free Sias due to the natural conditions for this reaction. The pyridoxamine dihydrochloride dose not interact with Sias at the carbon 2- position unless Sias is cleaved from sialoglycoconjugates in acidic conditions or by enzymatic action. The acetylacetone/ periodic acid method was reported in 1982 [301], the oxidation of bound Sias with periodic acid to form formaldehyde, which then reacts with an acetylacetone solution (200 μL acetylacetone, 15 g ammonium acetate, 300 μL glacial acetic acid, 99.5 mL water) at 60 °C for 10 min for the fluorescent formation at ex/em=410/510 nm [301]. The acetoacetanilide/ ammonia method is the most recently developed method, which was reported in 2008. The Sias is oxidized with the periodic acid to form the formaldehyde. The extra periodic acid needs to be neutralized by sodium thiosulfate. The fluorophore (ex/em=388/471) is formed by reacting with acetanilide and ammonium acetate at room temperature for 10 min [302]. This is the most sensitive method, which could measure Sia as low as 1 μg.

1.5.3 *Enzymatic methods for the analysis of Sias*

Enzymatic methods have been used to measure sialic acid, which employ complementing the colorimetric and fluorometric detection. These methods generally involve the cleavage of Sias from sialoglycoconjugates, following by transforming the free Sias into other compounds such as pyruvate and hydrogen peroxide, which are measured by colorimetric or fluorometric methods. The N-Acetyl-D-mannosamine was the first enzymatic method reported in 1962 [303]. The enzyme used converted Sias to form N-Acetyl-D-mannosamine and pyruvate. The N-Acetyl-D-mannosamine reacted with sodium tetraborate and 4-dimethyla-minobenzaldehyde to form the chromophore that absorption at 582 nm, which was used to quantify the Sias level in the sample. The pyruvate reacts with lactate dehydrogenase to form NADH measured photometrically at 340 nm [303].

The pyruvate/hydrogen peroxide method is another enzymatic method used for Sias quantification. Sias is converted to pyruvate and N-acetyl-D-mannosaime by Sia aldolase.

The N-acetyl-D-mannosaime will be converted to N-acetyl-D-glucosamine by N-acylglucosamine epimerase and then converted to hydrogen peroxide through N-acetyl hexosamine oxidase and peroxidase [304]. The hydrogen peroxide is then converted to the colorimetric substrate by peroxidase [305]. The *V. cholerae* sialidase/resorcinol is an enzymatic method coupled with a colorimetric detection. The *V. cholerae* sialidase is used to remove the terminal α-2,8 linkage Sias from sialoglycoconjugates [306]. Then the free Sias is quantified by the resorcinol method.

1.5.4 *Chromatographic and mass spectrometric assays for the determination of Sias*

Chromatographic and mass spectrometric methods are the most recently developed methods for the Sias qualification [101, 307-311]. The chromatography is required to partially separate Sias from other compound in complex bulk biological samples. Sias could be detected by Mass spectroscopy (MS) directly, or the derivatization of Sias could be measured through UV, fluorescence, or MS. Sias was determined by gas chromatography (GC), GC-MS, and high-performance liquid chromatography (HPLC)-UV [312, 313]. However, the GC and HPLC-UV have low sensitivity, which required large quantities of standard samples. The GC-MS method has a higher sensitivity to compared GC or HPLC-UV, which can measure as low as 20 pmol, although the methods are quite complicated. The other method was used to measure Sias in the biological samples involves the hydrolysis Sias from sialoglycoconjugates followed with active fluorescence substrate labeling. The fluorescence Sias derivatized substrate could be detected coupled with the HPLC separation. There are few active fluorescence substrates that have been reported recently. 4-Hydrazino-2-stilbazole is one of the reagents for α-keto acids, giving a fluorescent compound with maximum emission at 550 nm. 0.1 mL of plasma hydrolyzed under the acidic condition at 75°C for 1 h and followed with incubated with 4-hydrazino-2-stilbazole in 0.5 M ammonium chloride solution at 85°C for 15 min in the dark [314]. The HPLC method was used to separated Sias substrate with other interference compounds in the plasma, but pyruvate and two other peaks was noted to interfere with the Sias' peak. The fluorophore has a short life, which can only exist for 2 h under the light. The second active fluorescence compound was malononitrile, the formed fluorophore with maximum emission at 430nm [315, 316]. A reverse-phase column was used to separate Sias and other compounds in plasma. Sias has a short retention time, which is the advantage of the reverse-phase column. The phenyldiamine·2HCl was the third active fluorescence substrate for Sias labeling to form the fluorophore at ex/em=232/420 nm [317]. However, this reaction takes an extremely long time as 48h at RT or 16h at 37 °C, and is subject to interference substrates formed could affect the quantitative analysis. More recently, the 1,2-Diamino-4,5-methyleneoxybenzene (DMB) have been used to label the Sias to form the fluorophore at ex/em=373/448 nm. Sias in the biological samples is derivatized with DMB dissolved in mercaptoethanol and sodium hydrosulfite buffer by heated at 50 °C for 2h [318]. Sias is hydrolyzed at 70-80 °C from the sialoglycoconjugates by low concentration acid, such as 25mM sulphuric acid [319, 320], 0.1M TFA [318], 1M formic acid [321], and 2 M acetic acid [322]. Like DMB, DMBA and DAT can also be used for Sias derivatization [323, 324]. The DMB method is the most specific and sensitive method for Sias detective as low as 12 pg. When the MS is coupled with the HPLC for Sias detection, this method could be used to detect the different species of Sias, either derivatized or non-derivatized [101, 323-325]. This relatively sensitive method is able to detect as low as 10ng Sias, more specificity as there are no interference considerations due to the efficient separation and detection by

MS. A tandem MS/MS method coupled with the HPLC was used for quantifying Sias [101, 326], Sias derivatives [323-325, 327], and Sia-containing oligo and polysaccharides [328-330] in biological samples. This method uses two MS in tandem. The first MS ionized the sample and separated ions by mass to charge (m/z) ratio. The second MS was used to energy to collision the sample to form the fragments. The third MS was used to fragment the selected analytes for the final fragmented Sias detection [101, 323-326, 331]. This method could identify the different Sias linkages by linkage-specific derivatization [332-334] through the Sia linkage-specific alkylamidation, in which only α-2,3 linkage Sias gives different masses based on the linkage.

CHAPTER II

SYSTEMIC PROFILING OF DESIALYLATION OF THP-1 MONOCYTES AND

MACROPHAGES UPON LPS STIMULATION

2.1 Introduction

Sialic acid (Sias) is one of the special monosaccharides commonly located at the terminal of glycoconjugates, which contain nine carbon. Sias attach galactose (Gal) through α-2,3 or α-2,6 linkage, N-acetylgalactosamine (GalNAc) or N-acetylglucosamine (GlcNAc) through α-2,6 linkage; or another Sia through α-2,8-linkage [10]. Specific linkages and modification of Sias are involved in the tissue-specific and developmentally regulated manner. Sias are involved in many pathological and physiological processes, including cell adhesion, migration, inflammation, tumor, and tumor progression [9].

Lipopolysaccharide (LPS) is the major component of the outer membrane of Gram-negative bacteria. LPS activates the innate immune response in humans by stimulating inflammatory cytokine release such as tumornecrosis factor-alpha (TNFa) [335]. Monocytes and macrophages are an essential component of innate immunity, which are members of the mononuclear phagocyte system [336]. Monocytes are bone marrow-derived leukocytes found in the blood and spleen, which could recognize the foreign antigens, secrete chemokines and cytokines through recognition receptors while infection or cell injure [336]. Monocytes could be able to differentiate into macrophages or dendritic cells once monocytes are recruited to tissues. Macrophages can phagocyte pathogens or toxins, secrete chemokines or cytokines to recruit other immune cells, and migrate to local lymph nodes for antigen processing [336]. THP-1 is a human monocytic cell line derived from an acute monocytic leukemia patient used as a model cell line for many types of research [337]. THP-1 monocytes could differentiate to macrophages via phorbol-12-myristate-13-acetate (PMA), the heterogeneity of macrophage, which was affected via the differentiation [338, 339]. In the past research, proteins and organelles are both changes via differentiation; proteins such as anti-apoptotic protein Mcl-1 and Toll-like receptors (TLR) and organelles which involve lysosomal and mitochondria [340, 341]. In the previous studies, the cell surface glycoconjugates were shown desialydation during differentiation [120, 342, 343], transformation [155, 344, 345] and LPS stimulated inflammation in eripheral blood mononuclear cells (PBMCs) [346].

Sias on the glycoconjugates could be removed by sialidase (neuraminidases, Neu). In mammalian cells, there are four sialidases involved in this process, which are Neu1, Neu2, Neu3, and Neu4 [57, 58]. Neu1 is the lysosomal sialidase, which was used to degrade the sialoglycoconjugates [59-61]; Neu2 is the cytosolic sialidase which highly involved in removing gangliosides in the cytosol [62, 63]; the membrane-associated Sialidase Neu3 is the enzyme located at the membrane and specific for gangliosides [64]; Neu4 was found on the outer membrane of mitochondria, which could work on a bunch of glycoconjugates including glycoproteins, gangliosides, and oligosaccharides [65]. Neu3 is the sialidase located on the plasma membrane. However, the lysosomal sialidase Neu1 was moved to the plasma membrane in several cell types in the innate immune system, such as macrophage [35], T lymphocytes [36], and myofibroblast [72]. The cell surface Neu1 has been shown to respond to the Toll-like receptor response to the antigens [104]. Recently, Neu1 was found contained on the surface of secreted exosome with LPS stimulation[347].

In this research, the THP-1 monocytes and macrophages were used to profile sialo-level change and sialidase Neu1 and Neu3 expression, relocation, and secretion during the inflammation. We hypothesis LPS stimulated inflammation could cause the desialydation of THP-1 macrophages, caused by the secretion and relocation of Neu1.

2.2 Experiments

2.2.1 *Reagents*

PMA (phorbol-12-myristate-13-acetate), and LPS were purchased from Sigma (St. Louis, Missouri). N-Acetyl-d-neuraminic acid-1,2,3-$^{13}C_3$ ($^{13}C_3$-Sia, 99%) were obtained from the Carbosynth US LLC. (San Diego, CA). Acetonitrile (HPLC grade), paraformaldehyde (PFA, 16% w/v) and acetic acid (ACS grade) were supplied by Fisher Scientific (Hanover Park, IL, USA). FITC-labeled Maackia Amurensis (MAA), FITC-labeled Sambucus Nigra (SNA) lectin, and FITC-labeled *Arachis hypogaea* (peanut) (PNA) lectin were obtained from bioWORLD (Dublin, OH). 4-Methylumbelliferyl N-acetyl-α-D-neuraminic acid (4-MU-NANA) sodium salt was purchased from Carbosynth LLC (San Diego, CA, USA). GM3 ganglioside was obtained from Avanti Polar Lipids (Alabaster, AL, USA). Rabbit Anti-Neu1 IgG (1 mg/mL), Rabbit Anti-Neu3 IgG (1 mg/mL), Goat anti-Rabbit IgG-HRP from MyBioSource (MyBioSource, Inc. San Diego, CA, USA). Goat anti-Rabbit IgG -Alx 647 from Abcam (Abcam, Inc. Cambridge, MA, USA). FITC-labeled Ovalbumin (OVA) was purchased from ThermoFisher Scientific (Grand Island, NY).

2.2.2 *Apparatus*

LC-MS/MS quantification was performed using a Nexera liquid chromatography system (Shimadzu, Columbia, MD, USA) coupled with a Qtrap 5500 mass spectrometer equipped with an electrospray ionization source (AB SCIEX, Framingham, MA, USA), and the data were analyzed using Analysis 1.6.1 software. A FACSCanto II system was used for the flow cytometry and data assessed by the BD FACSDiva software (BD Bioscience, Mountain View, CA). ECOMAX X-Ray Film processor (PROTEC GmbH & Co. KG). Tomy MX-305 High-Speed Micro Centrifuge (BioSurplus, Inc. CA. USA).

2.2.3 *Cell culture*

THP-1 monocytes (ATCC® TIB-202™) obtained from ATCC were cultured in RPMI 1640 medium supplemented with 10% fetal bovine serum (FBS; Gibco, Rockville, MD) and 1% penicillin/ streptomycin (P/S; Gibco, Rockville, MD) at 37 °C with 5% CO_2. THP-1 macrophages were differentiated from THP-1 monocytes by treating with PMA (10 ng/mL) for 48 h with RPMI 1640 medium containing 10% heat-inactive FBS and 1% P/S, at 37 °C with 5% CO_2. THP-1 monocytes and macrophages were treated for 72 h with 100 ng/mL LPS alone (constituted in 1× phosphate-buffered saline (PBS)).

2.2.4 *Plasma membaren purification*

Polylysine-coated cell tissue culture plates were prepared by incubating with 5mg/50mL polylysine solution (1mL/25cm^2 culture surface) for 5 minutes. Then the coated plates were rinsed with DI-water 3 times and dried in the cell culture hood for 2 hours [29]. Ten million cells were blow off the plates and placed on polylysine-coated tissue culture surfaces for 4 hours at 37 ℃, and cells were lysate with ice-cold DI water. Intracellular debris was removed with ice-cold TBS buffer washes [30]. The plasma membrane on the polylysine-colted cell tissue culture was removed and collected in 1 mL PBS; the membrane fraction was pelleted by centrifugation at 10,000×g for 15 min [31] and resuspended into 50μL PBS. Protein concentrations were measured by the Bradford method using a protein assay kit.

Ten million THP-1 macrophage cells were lysate with ice-cold DI water, and intracellular debris was removed with ice-cold TBS buffer washes [348]. The plasma membrane on the cell tissue culture was removed and collected in 1 mL PBS, the membrane fraction was pelleted by centrifugation at 10,000×g for 15 min [349], and resuspended into 50μL PBS. Protein concentrations were measured by the Bradford method using a protein assay kit.

2.2.5 *Enzyme amount detect by Western Blotting*

A million cells in the indicated conditions were lysed with a mixture of RIPA buffer and protease inhibitor cocktail (1:100, v/v) on ice for 5 min, and the cell lysates were collected and clarified by centrifugation at 15,000 g at 4 °C for 10 min. Protein concentrations were measured by the Bradford method using a protein assay kit. Proteins at 30 μg/sample were separated on SDS-12 % polyacrylamide gels and then transferred to polyvinylidene difluoride membrane. The membranes were blocked with 5 % nonfat milk for 1 h at room temperature, followed by incubation with primary antibodies at 4 °C overnight. After washing with tris-buffered saline containing 0.05 % Tween 20 (TBS-T) buffer, the membrane was probed with specific secondary antibodies conjugated with horseradish peroxidase for 1 h at room temperature. Protein detection was accomplished by using the ECL Prime kit (Pierce) with the ECOMAX X-Ray Film processor.

2.2.6 *Enzyme activity*

The cell culture medium was collected from monocyte and macrophage with or without LPS stimulation. The activity of Neu1 and Neu3 in the cell culture medium was determined by mixing 10 μL sample into 0.05 M sodium acetate at pH 4.5 or 0.1 % Triton X-100, and 0.1 % BSA of PBS at pH 7.4, which contains 0.125 mM 4-MU-NANA and 0.250 mM ganglioside GM3 with or without 10μM Sialidase inhibitor 2,3-dehydro-2-deoxy-N-acetylneuraminic acid (DANA), respectively [29-31].

The cell lysate or isolated plasma membrane was resuspended in the reaction buffer containing 0.05 M sodium acetate at pH 4.5, 0.1 % Triton X-100, and 0.1 % BSA or pH 7.4, PBS. The activity of Neu1 and Neu3 was determined by the addition of 0.125 mM 4-MU-NANA and 0.250 mM ganglioside GM3 with or without 10μM Sialidase inhibitor 2,3-dehydro-2-deoxy-N-acetylneuraminic acid (DANA), respectively [29-31]. After 0.5 h incubation at 37 °C, the reaction mixtures were centrifuged to remove cellular debris, and the supernatants were precipitated with acetonitrile at a 1:3 ratio to remove the soluble proteins. Then 20 μL of the supernatant was mixed with 5 μL of internal standard (IS) and 75 μL of PBS, and 5 μL of the mixture was subjected to the LC-MS/MS analysis. Protein concentrations in each assay were measured by the Bradford method using a protein assay kit (Bio-Rad, Hercules, CA, USA). One unit of sialidase activity was defined as 1 nmol of free Sia that was released from 1 mg of total protein per hour at 37 °C.

2.2.7 *LC-MS/MS sample preparation*

To quantify the total Sia of the cell, 10^6 cells/mL were collected and ultra-sonicated to yield a homogeneous mixture, then the cell lysis was hydrolyzed with 2 M acetic acid (1:1) at 80 °C for 90 min, then 5 µL of the internal standard (IS) stock solution (500ng/mL) and 95 µL of the lysis sample was mixed and 5 µL of the mixture was injected to LC-MS/MS.

2.2.8 *LC-MS/MS analysis*

The Primesep D column (2.1 × 100 mm, 5 µm; SIELC Technologies, Prospect Heights, IL, USA) was used to separate the Neu5Ac, Neu5Gc, and IS. The binary gradient was used in the HPLC, in which phase A was deionized water with 10 mM ammonium formate and 0.1% formic acid, and phase B was 90% acetonitrile and 10% water with 3 mM ammonium formate and 0.04% formic acid. The column was equilibrated and started with 10% phase B, which lasted for 1 min, then the gradient was changed to 95% B within 3 min and kept at 95% B for 2 min. Finally, the mobile phase was changed to 10% B in 0.1 min and maintained for 2.9 min. For each run of the samples, a total 9 min were required. To qualitatively measure the amount of Sia, the MRM transitions were set at m/z 308.2→87.0 for Neu5Ac, 324.0→116.0 for Neu5Gc, and 311.2→ 89.10 for IS $^{13}C_3$-Neu5Ac.

2.2.9 *Method validation and sample preparation*

Sia calibration standard stock solution was prepared in PBS at concentrations of 20, 50, 150, 500, 1500, 4500, 10000 ng/mL. Quality control (QC) samples of stock solutions were prepared at low (40 ng/mL), mid (800 ng/mL), and high (8000 ng/mL) concentrations under the same conditions. IS ($^{13}C_3$-Sia) stock solution was prepared in PBS at 500 ng/mL. To prepare the sample, internal standard stock solution (5 µL), standard or QC stock solution (5 µL), and mobile phase A (90 µL) were mixed well, and 5 µL of the mixture was injected into the LC-MS/MS for analyzing.

2.2.10 *Flow cytometry*

Flow cytometry was used to determine the cell surface Sia linkage; THP-1-derived macrophages (100,0000 cells) were treated with LPS (100 ng/mL). The treated cells were collected by blow off the plates and washed with PBS, cell pellet suspended in FITC-MAA (20 μg/mL), FITC-SNA (20 μg/mL) or PNA (20 μg/mL), and incubated at room temperature for 30 min, washed with cold PBS for 2-3 times, and finally resuspended in PBS with 0.5% sodium azide. 10,000 cells per condition were tested each time.

2.2.11 *Statistical analysis*

Unless otherwise noted, all data was represented as average ± standard error (SE) from n=3 wells/conditions, with at least three independent repeats of each assay. Significance in differences between groups was analyzed using two-way analysis of variance (ANOVA) with Tukey post-hoc HSD test, and p-values < 0.05 were considered statistically significant.

2.3 Results and Discussion

2.3.1 *The desialylation of THP-1 monocytes and macrophages stimulated by LPS*

LPS stimulation of monocyte and macrophage could cause endogenous sialidase expression, which causes desialylation that enhances secretion of the pro-inflammatory cytokine, including TNF-α and IL-6 [350, 351]. In order to examine the desialylation extent, the total Sia level in cell lysates of THP-1 monocytes and macrophages upon LPS stimulation were quantified by LC-MS/MS method, respectively. As a result, after 24 h LPS stimulation (100 ng/mL), the total Sia levels decreased with p<0.1 in the cell lysates of both THP-1 monocytes and macrophages (Figure 2. 1A and 1B), indicating that desialydation of the glycoconjugates of THP-1 monocyte and macrophage occurred during LPS stimulation. Accordingly, the free Sias cleaved from glycoconjugates could be released into the cell culture medium. Therefore, free Sias in the cell culture medium in THP-1 monocytes and macrophages treated with and without LPS were quantified by LC-MS/MS. As shown in Figure 2. 2A, the free Sia significantly increased in the culture medium of THP-1 monocytes upon LPS stimulation. However, no apparent changes in the medium of macrophages after LPS stimulation (Figure 2. 2B). The reduction of cell surface Sia could contribute to the level of free sialic acid in the cell culture medium. However, the free Sia in the cell culture medium could be metabolized, such as by oxidation, and even be re-taken by the cells due to insufficient biosynthesis under cell activation. Further, comparing with THP-1 monocytes (Figure 2. 2C), the total Sia, including free Sia and Sia released from sialoglycoconjugates by hydrolysis with 1M acetic acid, was found to increase significantly in the THP-1 macrophages upon LPS stimulation (Figure 2. 2D), indicating that there might be Sia-containing glycoproteins secreted from THP-1 macrophages upon LPS stimulation.

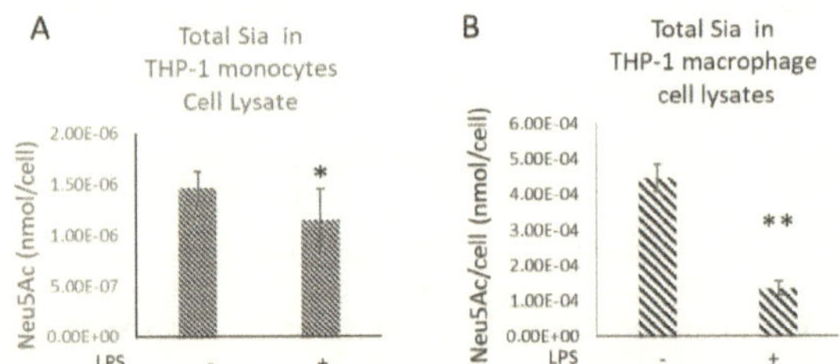

Figure 2. 1 Total Sialic acid in THP-1 monocyte and macrophage upon LPS stimulation. **Indicates p < 0.05 between the groups, *Indicates p<0.1 between the group. Data were presented as average ± SE (n = 3 wells/condition), with three independent repeats of the experiment.

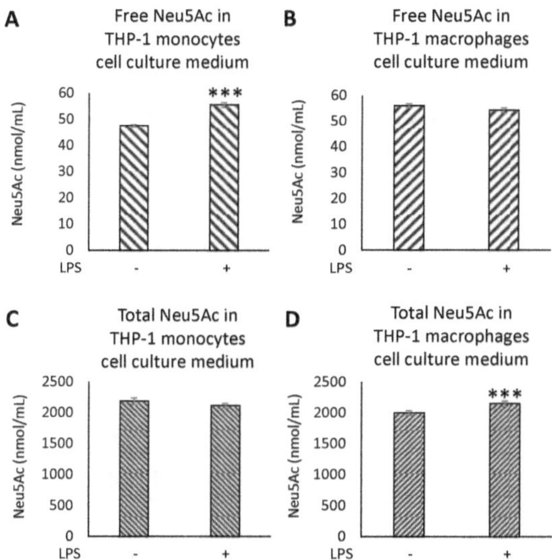

Figure 2. 2 Sia in the cell culture medium of THP-1 monocytes and macrophages upon LPS stimulation (100ng/mL, 24 h). (**A**) Free SA in the cell culture medium of THP-1 monocytes upon LPS stimulation; (**B**) Free SA in the cell culture medium of THP-1 macrophages upon LPS stimulation; (**C**) Total SA in the cell culture medium of THP-1 monocytes upon LPS stimulation; (**D**) Total SAin cell culture medium of THP-1 macrophages upon LPS stimulation. ***Indicates p < 0.01 between the groups. Data were presented as average ± SE (n = 3 wells/condition), with three independent repeats of the experiment.

2.3.2 *THP-1 monocyte and macrophage cell surface linkage change upon LPS*

stimulation

Sias attach to galactose (Gal) or N-acetylgalactosamine (GalNAc) through α2,3- or α2,6-linkage (Varki 2008). Desialylation could cause exposure of Gal or GalNAc on the cell surface glycoproteins. Therefore, we examined cell surface Gal level changes of THP-1 monocytes and macrophages treated with and without LPS with FITC-labeled Arachis hypogaea (peanut) (PNA), which explicitly binds to the terminal β-galactose. As a result, the Gal level significantly increased on the cell surface of macrophages upon LPS stimulation (Figure 2. 3B). However, Gal level on THP-1 monocytes did not change (Figure 2. 3A). These results indicate a different desialylation pattern between monocytes and macrophages, which deserves further study in the future.

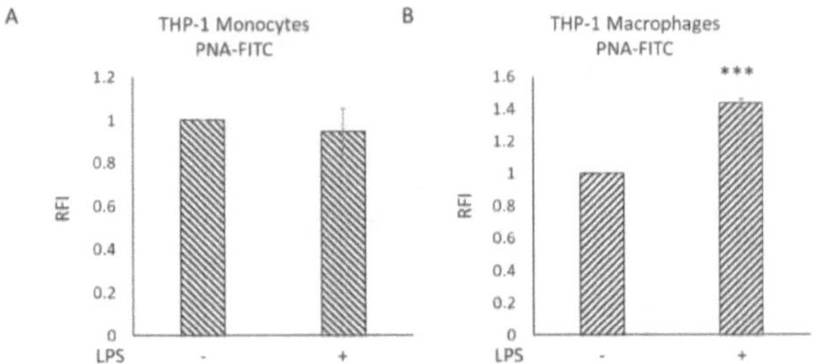

Figure 2. 3 Flow cytometry examination of cell surface Gal of THP-1 monocytes and macrophages upon LPS stimulation. (A) PNA-FITC binding of Gal on THP-1 monocytes; (B) PNA-FITC binding of Gal on THP-1 macrophage. ***Indicates p < 0.001 between the groups. Data were presented as average ± SE (n = 3 wells/condition), with three independent repeats of the experiment.

2.3.3 *Sialidase activities on THP-1 macrophage cell lysates, plasma membrane, and cell culture medium upon LPS stimulated inflammation.*

LPS stimulation could cause endogenous sialidase expression that catalyzes the removal of Sia from both monocytes and macrophages. In this study, the sialidase activity of Neu1 and Neu3 in THP-1 monocytes and THP-1 macrophages were determined by using the exogenous sialidase substrate 4-methylumbelliferyl N-acetyl-α-D-neuraminic acid (4-MU-Neu5Ac) and ganglioside GM3, respectively. Specifically, Neu1 and Neu3 sialidases activity of cell lysates, cell plasma membrane, and cell culture medium of THP-1 monocytes and THP-1 macrophages treated with and without LPS were quantified by measuring the released free Sia by LC-MS/MS method. Neu1 activity was examined on 4-MU-NANA at pH 4.5 and 7.4, respectively, by following the literature method [280]. Neu3 activity for a ganglioside substrate (GM3) was examined at pH 4.5 as described in the literature [316]. As a result, Neu1 activity increased 3.1 folds at pH 7.4 and 1.3 folds at pH 4.5 in the cell lysates (Figure 2. 4A and 4B), but decreased 0.24 folds at pH 7.4 and 0.40 folds at pH 4.5 in the cell plasma membrane (Figure 2. 4D and 4E) of THP-1 monocytes upon LPS stimulation. The changed activity of Neu1 in the cell plasma membrane might be due to its secretion from the cell plasma membrane. As shown in Figures 4G and 4H, Neu1 activity increased 1.3 fold at pH 7.4 and 1.4 fold at pH 4.5 in the cell culture medium of THP-1 monocytes upon LPS stimulation. In addition, it may be due to the incomplete formation of the matured PPCA-Neu1 heterodimer on the plasma membrane [82]. In comparison, Neu3 activity increased 0.68 folds in cell lysates (Figure 2. 4C) and 0.17 folds with p<0.1 value (Figure 2. 4F), indicating the upregulation of Neu3 in the cell plasma membrane of THP-1 monocytes upon LPS stimulation. Further, Neu3 activity increased 2.3 fold in the cell culture medium of THP-1 monocytes upon LPS stimulation (Figure 2. 4I). These results firmly indicate that THP-1 monocytes highly express both Neu1 and Neu3 and even secrete Neu1 and Neu3 as well upon LPS stimulation. We speculate that both cell surface Neu1 and secreted Neu1 could desialylate cell surface TLR4 and contribute to LPS/TLR4 pathway in monocytes. The desialylation capacity of secreted Neu1 from monocytes will be examined in our continuing study.

Next, we quantified Neu1 and Neu3 sialidase activity of cell lysates and cell plasma membrane of macrophages treated with and without LPS (100 ng/mL for 24 h) with the exogenous sialidase substrates 4-MU-Neu5Ac and GM3, respectively. As a result, Neu1 activity increased 1.1 fold at pH 7.4 and 1.3 fold at pH 4.5 in cell lysates (Figure 2. 5A and 5B) and 3.8 fold at pH 7.4 and 4.5 fold at pH 4.5 on cell plasma membrane (Figure 2. 5D and 5E) of THP-1 macrophages upon LPS stimulation. In comparison, Neu3 activity increased 1.4 fold in cell lysates (Figure 2. 5C) and increased 15 fold on the cell plasma membrane (Figure 2. 5F) of THP-1 macrophages upon LPS stimulation. Finally, we quantified Neu1 and Neu3 sialidase activity in the cell culture medium of macrophages treated with and without LPS (100 ng/mL for 24 h). As a result, the Neu1 activity increased about 1.6 fold at pH 7.4, but only 1.04 fold increased at pH 4.5, and Neu3 activity increased 2.0 folds in the cell culture medium of macrophage upon LPS stimulation (Figure 2. 5G, 5H, and 5I). These results firmly indicated that LPS induces higher Neu1 and Neu3 expression and secretion from macrophages as well. We speculate that both cell surface Neu1 and secreted Neu1 will desialylate cell surface TLR4 and contribute to LPS/TLR4 pathway in macrophages. The desialylation capacity of secreted Neu1 from macrophages will be examined in our continuing study.

Multiple enzymes exist in the cell culture medium, which may also cause hydrolysis of the sialidase substrate 4-MU-Neu5Ac and GM3 [352]. In this study, sialidase inhibitor DANA was used to confirm the specific sialidase activity in the cell culture media. As a result, DANA inhibited both Neu1 and Neu3 activities in the cell culture medium of macrophages (Figure 2. 6A, 6B, and 6C) and Neu1 and Neu3 activity at pH 4.5 in the cell culture medium of THP-1 monocytes macrophages (Figure2. 6E and 6F). However, enhanced activity was detected for Neu1 activity at pH 7.4 in THP-1 monocytes (Figure 2. 6D). The enhanced activity might be caused by unknown glycosidases at pH 7.4. Nevertheless, the Neu1 sialidase activity at pH 7.4 is only around 5% of the activity at pH 4.5 in the cell culture medium of the THP-1 monocytes (Figure 2. 4 G and H). Interestingly, the Neu1 sialidase activity is about 45 folds enhanced in the cell culture medium of THP-1 macrophages compared to the THP-1 monocytes (Figure 2. 5G and Figure 5G).

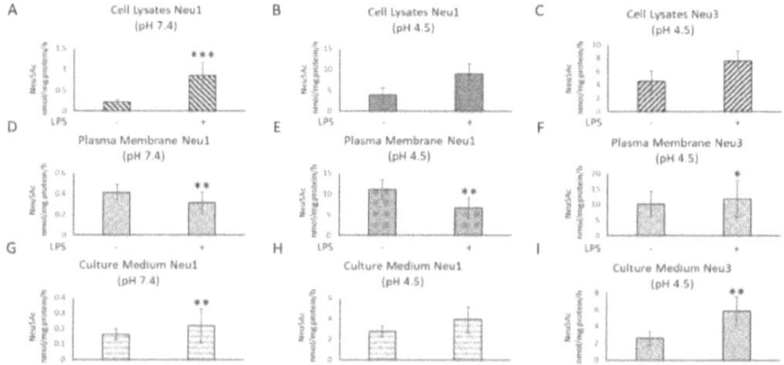

Figure 2. 4 Sialidase activity of THP-1 monocytes upon LPS stimulation. **(A)** Neu1 activity at pH 7.4 in cell lysates; **(B)** Neu1 activity at pH 4.5 in cell lysates; **(C)** Neu3 activity in cell lysates; **(D)** Neu1 activity at pH 7.4 on cell plasma membrane; **(E)** Neu1 activity at pH 4.5 on cell plasma membrane; **(F)** Neu3 activity on cell plasma membrane; **(G)** Neu1 activity at pH 7.4 in cell culture medium; **(H)** Neu1 activity at pH 4.5 in cell culture medium; **(I)** Neu3 activity in cell culture medium. * Indicates $p < 0.1$, **Indicates $p < 0.05$, ***Indicates $p < 0.001$ between the groups. Data were presented as average $\pm$ SE (n = 3 wells/condition), with three independent repeats of the experiment. Data were presented as average $\pm$ SE (n = 3 wells/condition), with three independent repeats of the experiment.

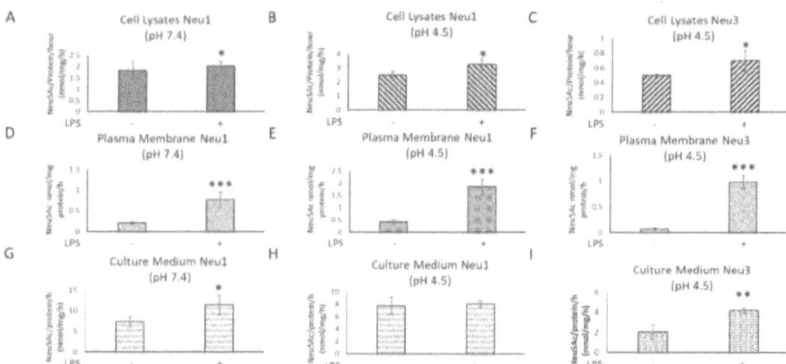

Figure 2. 5 Sialidase activity in THP-1 macrophages upon LPS stimulation. **(A)** Neu1 activity at pH 7.4 in cell lysates; **(B)** Neu1 activity at p H4.5 in cell lysates; **(C)** Neu3 activity in cell lysates; **(D)** Neu1 activity at pH 7.4 on cell plasma membrane; **(E)** Neu1 activity at pH 4.5 on cell plasma membrane; **(F)** Neu3 activity on cell plasma membrane; **(G)** Neu1 activity at pH 7.4 in cell culture medium; **(H)** Neu1 activity at pH 4.5 in cell culture medium; **(I)** Neu3 activity in cell culture medium. * Indicates $p < 0.1$, **Indicates $p < 0.05$, ***Indicates $p < 0.001$ between the groups. Data were presented as average $\pm$ SE (n = 3 wells/condition), with three independent repeats of the experiment. Data were presented as average $\pm$ SE (n = 3 wells/condition), with three independent repeats of the experiment.

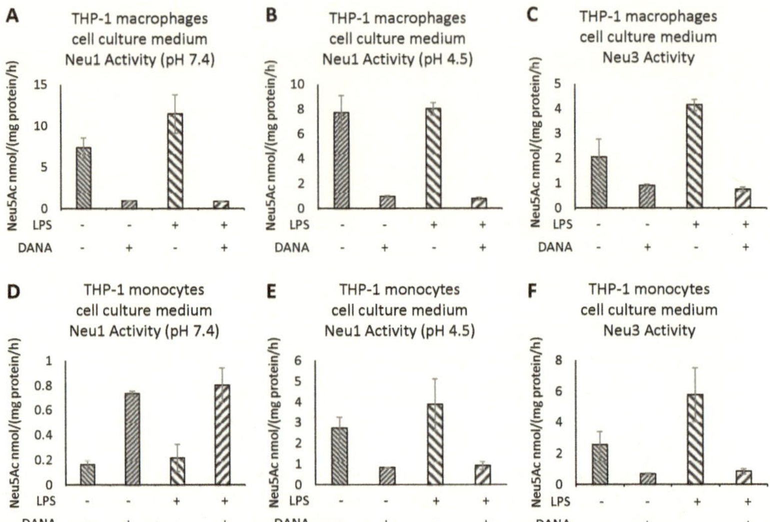

Figure 2. 6 Sialidase activity in THP-1 monocyte and macrophage cell culture medium upon LPS stimulation. **(A)** Neu1 activity at pH7.4 in THP-1 macrophage cell culture medium with or without sialidase inhibitor DANA; **(B)** Neu1 activity at pH4.5 in THP-1 macrophage cell culture medium with or without sialidase inhibitor DANA; **(C)** Neu3 activity in THP-1 macrophage cell culture medium with or without sialidase inhibitor DANA; **(D)** Neu1 activity at pH7.4 in THP-1 monocyte cell culture medium with or without sialidase inhibitor DANA; **(E)** Neu1 activity at pH4.5 in THP-1 monocyte cell culture medium with or without sialidase inhibitor DANA; **(F)** Neu3 activity in THP-1 monocyte cell culture medium with or without sialidase inhibitor DANA.

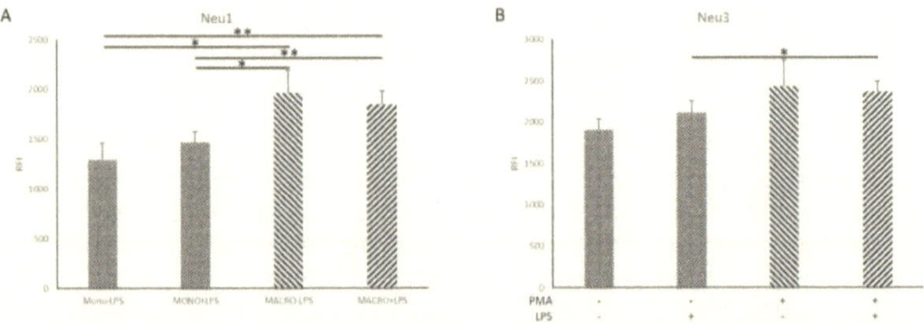

Figure 2. 7 Sialidase expression on THP-1 monocytes and macrophages upon LPS stimulation. **(A)** Neu1; **(B)** Neu3. **Indicates $p < 0.05$, * indicates $p<0.1$ between the groups. Data were presented as average ± SE (n = 3 wells/condition), with three independent repeats of the experiment.

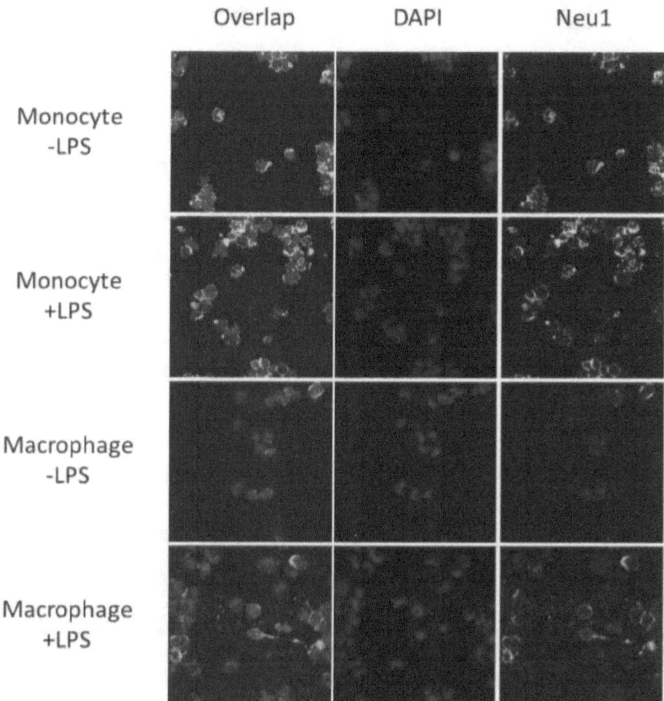

Figure 2. 8 Neu1localization in THP-1 monocyte and macrophage cells. The primary antibody are IgG Rabbit anti-Neu. Secondary Antibody is Cy5 labeled anti-Rabbit IgG (yellow), and nucleus are stained by DAPI (Blue).

2.3.4 *The sialidase expression on THP-1 monocyte and macrophage upon LPS stimulation*

Flow cytometry (Figure 2. 7) was used to detect sialidase Neu1 and Neu3 on the cell surface, respectively. Meanwhile, the location of Neu1 was detected by confocal microscopy (Figure 2. 8). As a result, Neu1 and Neu3 were not changed in both THP-1 monocyte and macrophage and cell surface upon LPS stimulation (Figure 2. 7). However, both expression levels of Neu1 and Neu3 are noticed during differentiation. Due to the western blot sensitivity issue, the sialidase amount could not be able to detect in the cell culture medium. A more sensitive method is required to be considered in the future.

2.4 Conclusion

This study characterized the sialylation status of LPS-stimulated THP-1 macrophages and sialidase expression and activities. We confirmed that the desialydation (Figure 2. 1) of LPS stimulated THP-1 macrophage is closely related to the enhanced sialidase Neu1 and Neu3 activity (Figure 2. 4, Figure 2. 5, and Figure 2. 6). However, Neu1 was not relocated from the lysosome to the plasma membrane under LPS stimulation. Many studies have shown the PMA caused Neu1 relocalization during the differentiation of monocytes to macrophages, and the sialidase Neu1 activity was enhanced by the PPCA[36, 70, 353]. Since the PMA induced differentiation of THP-1monocytes already caused the relocalization of Neu1 through a similar pathway to the LPS stimulation, which caused the second time LPS stimulation, the Neu1 relocalization on THP-1 macrophage could be detected. However, the second time LPS stimulation may cause the sialidase secretion to the surrounding environment. The concentration of Neu1 in the cell culture medium could not be detected due to the Western blotting detection limitation, but the activity of Neu1 and Neu3 are enhanced in the cell culture medium. However, the activities of Neu1 and Neu3 are enhanced in the THP-1 macrophage cell culture medium (Figure 2. 5 and 6 G, H, and I, and Figure 2. 7).

In conclusion, the desialylation profile of THP-1 monocytes and THP-1 macrophages upon LPS stimulation were systemically examined. We confirmed that the desialylation of LPS stimulated THP-1 monocytes and THP-1 macrophages are closely related to the enhanced sialidase Neu1 and Neu3 activity. In addition, we found that enhanced Neu1 and

Neu3 sialidase activities in the cell culture medium, which support Neu1 and Neu3 release from THP-1 monocytes and THP-1 macrophages upon LPS stimulation. The secreted sialidase could serve as exogenous sialidase to desialylate cell surface glycoconjugates as well. This is the first report about sialidase Neu1 and Neu3 expression, relocation, and secretion from monocytes and macrophages upon LPS stimulation, indicating that endogenous sialidase expression, relocation, and secretion are all involved in LPS/TLR4 signaling pathway and subsequent biological processes and cellular functions.

CHAPTER III

CHEMICAL MODULATION OF DESIALYLATION OF MACROPHAGES AND

THEIR PROPERTY AND FUNCTION

3.1 Introduction

Sialic acids (Sias) are a family of 9-carbon containing acidic monosaccharides located at the terminal of glycan; based on the diversity on C-4, 5, 7, 8, and 9, more than 50 species Sias were found. However, the two most famous Sias with an amino function group at C-5 are either acetylated (N-acetylneuraminic acid, Neu5Ac) or glycolylated (N-glycolylneuraminic acid, Neu5Gc). They attached galactose (Gal) or N-acetylgalactosamine (GalNAc) units via α2,3- or α2,6-linkage, or another Sia via α2,8-linkage on both N- and O-linked glycans [354]. Given their terminal location on glycans and hydrophilic and electronegative features, cell surface Sias contribute to morphological and functional characteristics of the cells and thus play important roles in human health and diseases [11]. First, Sias has structural and modulatory roles due to their negative charge and hydrophilicity. Second, Sias serve as components of binding sites for many receptors and various pathogens [10]. Macrophages play an essential role in host defense from infection, control of inflammation, and injury repair [355-357]. An effective immune response of macrophages is necessary for maintaining homeostasis and health in the human body. Cell surface receptors and secreted functional molecules are the key modulators of

50

macrophage functions. In addition, cellular mechanics such as elasticity is also a significant determinant of macrophage function as cellular elasticity may affect cell surface receptor functions [358]. It is well recognized that biologic modulators such as lipopolysaccharide (LPS) or cytokines modulate cell elasticity and functions of macrophages [326, 359, 360]. For example, phagocytosis is an important function of macrophages, which requires macrophages to exert physical forces on particles, and thus macrophage elasticity may affect phagocytosis [361-363]. LPS stimulates phagocytosis and increases macrophage cellular elasticity [364]. The dense layer of glycans on the macrophage cell surface is involved in both functions and properties, including elasticity. The effect of polysialic acid on isogenic C6 rat glioma cells was quantified using AFM; cells overexpressing polysialic acid was shown more compliance than wild-type cells [365]. Despite these efforts, sialylation status on the biophysical properties of macrophages remains unexplored.

Recent advances in soft matter physics and mechanical engineering paved the way for understanding cellular mechanics by applying various experimental and theoretical models. Micropipette aspiration, optical tweezers, magnetic twisting cytometry, atomic force microscopy (AFM), to name a few, have been used for mechanical characterization of cells and tissues. Among them, AFM offers a unique advantage to obtain the force measurements at physiologically relevant conditions and determine various mechanical properties [366, 367]. Changes in the cellular microenvironment such as pathogens and inflammatory molecules could disrupt cell phenotype and modulate cell mechanical properties, which could be measured reliably by AFM. AFM utilizes a microfabricated cantilever tip for the mechanical characterization of cells and tissues by generating force-distance curves. For instance, the mechanism of cell stiffening for cytokine release was assessed in macrophage differentiated THP-1 cells by subjecting to AFM analysis with PRDX5-functionalized cantilever tips [367]. The changes in mechanical properties of LPS-stimulated macrophages were previously quantified, although such measurements were limited to quantification of elastic modulus and adhesion13. Colloidal force microscopy was also used to calculate the mechanical properties of LPS-stimulated macrophages, but the force measurements were not performed under physiological conditions [364].

The removal of Sias (desialylation), an essential part of the Sia metabolism, is of axiomatic importance in many biological processes [368]. For example, LPS induces up-regulation of endogenous sialidases, which cause desialylation and significantly enhances the secretion of pro-inflammatory cytokines such as TNF-α and IL-6 [350, 351]. There is a good understanding of the role of desialylation in LPS-induced activation of macrophages; however, the effect of sialylation and desialylation on cellular elasticity has not been explored. In this report, we quantitatively investigated the sialylation status of Human THP-1 monocyte-derived macrophage upon LPS stimulation and explored the metabolic modulation of the cell sialylation status by culturing the cells with free sialic acids (Neu5Ac, Neu5Gc) and sialidase inhibitors. Human THP-1 macrophage has been a standard model to study monocyte/macrophage transition, functions, mechanisms, and signaling pathways and present relatively similar response patterns as human peripheral blood mononuclear cells [369]. With the sialylation status determined at these conditions, we quantified the cellular elastic properties and function of the THP-1 macrophages. Overall, a systematic recognition of sialylation status related to the cellular elasticity of macrophages will contribute to defining macrophage phenotypes and understand macrophage functional diversity. Further, it will provide novel mechanisms and approaches for disease diagnosis and treatment.

3.2 Experiments

3.2.1 *Reagents*

Neu5Ac2en (N-Acetyl-2,3-dehydro-2-deoxyneuraminic acid), LPS (Lipopolysaccharide, from E. coli O55:B5), Oseltamivir phosphate, and PMA (phorbol-12-myristate-13-acetate) were purchased from Sigma (St. Louis, Missouri). Neu5Ac (N-Acetyl-D-neuraminic acid) was purchased from Rose Scientific Ltd (Edmonton AB, Canada). N-Acetyl-D-neuraminic acid-1,2,3-$^{13}C_3$ ($^{13}C_3$-Sia, 99%) and N-glycolyl neuraminic acid (Neu5Gc) were purchased from the Carbosynth US LLC. (San Diego, CA). Acetonitrile (HPLC grade), paraformaldehyde (PFA, 16% w/v), and acetic acid (ACS grade) were purchased from Fisher Scientific (Hanover Park, IL, USA). FITC-labeled Maackia Amurensis (MAA) and FITC-labeled Sambucus Nigra (SNA) lectin were purchased from bioWORLD (Dublin, OH). FITC-labeled Ovalbumin (OVA) was purchased from ThermoFisher Scientific (Grand Island, NY).

3.2.2 *Apparatus*

LC-MS/MS quantification was performed using a Nexera liquid chromatography system (Shimadzu, Columbia, MD, USA) coupled with a Qtrap 5500 mass spectrometer equipped with an electrospray ionization source (AB SCIEX, Framingham, MA, USA), and the data were analyzed using Analysis 1.6.1 software. A FACSCanto II system was used for the flow cytometry and data assessed by the BD FACSDiva software (BD Bioscience, Mountain View, CA). An MFP-3D-Bio atomic force microscope (AFM; Oxford Instruments; formerly Asylum Research, Santa Barbara, CA) mounted on an inverted epi-fluorescence microscope (Nikon Eclipse Ti) was used for live cell nano-indentation assay using TR 400, PSA short lever cantilevers (nominal spring constant ~ 0.02 N/m).

3.2.3 *Cell culture*

THP-1 monocytes (ATCC® TIB-202™) obtained from ATCC were cultured in RPMI 1640 medium supplemented with 10% fetal bovine serum (FBS; Gibco, Rockville, MD) and 1% penicillin/ streptomycin (P/S; Gibco, Rockville, MD) at 37 °C with 5% CO_2. THP-1 macrophages were differentiated from THP-1 monocytes by treating with PMA (10 ng/mL) for 48 h with RPMI 1640 medium containing 10% heat-inactive FBS and 1% P/S, at 37 °C with 5% CO_2. THP-1 macrophages were treated with 500 μM Neu5Ac, 500 μM Neu5Gc, 2 mM Neu5Ac2en, and 9.4 μM Oseltamivir phosphate (constituted in 1× phosphate-buffered saline (PBS) for 72 h, respectively. In separate cultures, THP-1 macrophages were treated with 100 ng/mL LPS with and without Neu5Ac, Neu5Gc oseltamivir phosphate, and Neu5Ac2en in the same concentrations for 72 h, respectively. Control culture received the same volume of PBS (pH7.4) without exogenous compounds.

3.2.4 *Cell viability assays*

To test THP-1 macrophage viability, 10,000 cells were treated with 500 μM Neu5Ac, 500 μM Neu5Gc, 2 mM Neu5Ac2en, 9.4 μM Oseltamivir phosphate (constituted in 1× phosphate-buffered saline (PBS) for 72 h, respectively. In separate cultures, THP-1 macrophages were treated with 100 ng/mL LPS with and without Neu5Ac, Neu5Gc, Neu5Ac2en, and oseltamivir phosphate in the same concentrations for 72 h, respectively. Control culture received the same volume of PBS without exogenous compounds. The mitochondria activity was assessed by incubation with 50 μg/mL MTT for 4 h, the cell medium was removed, and 150 μL of DMSO was added to each well, and the absorbance at 570 nm was read by an Epoch™ microplate spectrophotometer (Bio-Tek).

3.2.5 *LC-MS/MS sample preparation*

To quantify the total Sia of cells, 10^6 cells/mL were collected and ultra-sonicated to yield a homogeneous mixture, and then the cell lysis was hydrolyzed with 2 M acetic acid (1:1) at 80 °C for 90 min, then 5 μL of the internal standard (IS) $^{13}C_3$-Neu5Ac stock solution (500ng/mL) and 95 μL of the lysis sample was mixed and 5 μL of the mixture was injected to LC-MS/MS.

3.2.6 *LC-MS/MS analysis*

The Primesep D column (2.1 × 100 mm, 5 μm; SIELC Technologies, Prospect Heights, IL, USA) was used to separate the Neu5Ac, Neu5Gc, and IS. The binary gradient was used in the HPLC, in which phase A was deionized water with 10 mM ammonium formate and 0.1% formic acid, and phase B was 90% acetonitrile and 10% water with 3 mM ammonium formate and 0.04% formic acid. The column was balanced and started with 10% phase B, which lasted for 1 min, then the gradient was changed to 95% B within 3 min and kept at 95% B for 2 min. Finally, the mobile phase was changed to 10% B in 0.1 min and maintained for 2.9 min. For each run of the samples, 9 min were required. To qualitatively measure the amount of Sia, the MRM transitions were set at m/z 308.2→87.0 for Neu5Ac, 324.0→116.0 for Neu5Gc, and 311.2→ 89.1 for IS, $^{13}C_3$-Neu5Ac.

3.2.7 *Method validation and sample preparation*

Sia calibration standard stock solution was prepared in PBS at concentrations of 20, 50, 150, 500, 1500, 4500, 10000 ng/mL. Quality control (QC) samples of stock solutions were prepared at low (40 ng/mL), mid (800 ng/mL), and high (8000 ng/mL) concentrations under the same conditions. Internal standard ($^{13}C_3$-Neu5Ac) stock solution was prepared in PBS at 500 ng/mL. To prepare the sample, internal standard stock solution (5 µL), standard or QC stock solution (5 µL), and mobile phase A (90 µL) were mixed well, and 5 µL of the mixture was injected into the LC-MS/MS for analyzing.

3.2.8 *Flow cytometry*

To determine the cell surface Sia, all cells treated as in procedures were collected by blow off the plates and washed with PBS (pH 7.4), cell pellets were suspended in FITC-MAA (20 µg/mL) or FITC-SNA (20 µg/mL), and incubated at room temperature for 30 min, washed with cold PBS (pH 7.4) for 3 times, and finally resuspended in PBS (pH 7.4) with 0.5% sodium azide. 10,000 cells per condition were tested each time, and flow cytometry was performed on the FACSCanto II system.

3.2.9 *Atomic force microscopy*

THP-1 monocytes (10,000) were seeded onto a cover glass for 48 h with media containing 10 ng/mL PMA. After 48 h, THP-1-derived macrophages were treated as in **2.3 Cell Culture** procedures. Then, each cover glass was attached to an AFM-specific 50-mm Petri dish. Prior to each experiment, the actual spring constant was calculated from a force-distance curve using a thermal calibration method in a clean petri-dish containing DMEM. The cantilever was submerged in media for at least 30 min to reach mechanical and thermal equilibrium before obtaining force-deflection curves. Using a petri-dish heater, cells were maintained at 37 °C, and at least 30 cells were analyzed for each treatment/experiment at approach/retraction velocity of 5 µm/sec. The cells were indented by applying up to 2 nN force, and the resulting force-indentation curves were analyzed using proprietary software (Igor Pro 6.37). Young's moduli (modulus of elasticity or stiffness) of the cells were determined using a Hertz model assuming Poisson's ratio of 0.5 and considering 15 nm cantilever tip radius and 500-650 nm indentation depth. Adhesion, defined as the amount of force required to separate cell surface and tip, was calculated directly from the force-indentation curves. From the retraction curves, the tether force (F_T) was calculated directly from the series of force steps observed. Tether radius (R_T), defined as the connection between cytoskeleton and plasma membrane, was calculated from $F_T \cong 2\pi K_B/R_T$, where K_B is the bending stiffness (0.1-0.3 pN.μm)[370, 371]. Finally, the force required to deform a membrane, known as the apparent membrane tension, was calculated as $T_M \cong F_T^2/8\pi^2 K_B$ [371].

3.2.10 *OVA antigen uptake assay*

To test the macrophage function, OVA antigen uptake assay was used. THP-1-derived macrophages (100,000 cells) stimulated with and without LPS (100 ng/mL) were incubated with Neu5Ac (500 μM), Neu5Gc (500 μM), Neu5Ac2en (2 mM), and Oseltamivir phosphate (9.4 μM) for 72 h, respectively. Control cells received the same volume of PBS with no LPS for 72 h. Then, FITC-OVA (10 μg/mL) was added and incubated with all cells at 37 °C for 30 min, respectively. The extra FITC-OVA was washed out with cold PBS (pH 7.4), the cells were collected and resuspended in PBS (pH 7.4) with 0.5% sodium azide, and flow cytometry was performed on the FACSCanto II system.

3.2.11 *Statistical analysis*

Unless otherwise noted, all data was represented as average ± standard error (SE) from n=3 wells/conditions, with at least three independent repeats of each assay. Significance in differences between groups was analyzed using two-way analysis of variance (ANOVA) with Tukey post-hoc HSD test, and p-values < 0.05 were considered statistically significant.

3.3 Results and Discussion

Accumulated evidence suggests that cell surface Sias can dramatically impact cell properties and represent different cellular statuses, indicating cell surface Sias might contribute to macrophage cell elasticity and functionality [372]. Since changes in the mechanical behavior of cells have been associated with a variety of pathologies, it is desirable to investigate sialylation status related to macrophage elasticity and functions and explore their functional involvement. AFM has become an essential tool in examining the mechanical properties of a wide range of cells, including macrophages. This study investigated the effect of sialylation and desialylation on cell elasticity using THP-1 macrophages in LPS-induced activation and metabolic modulation.

3.3.1 *Cellular sialylation levels of THP-1 macrophages upon LPS stimulation and metabolic modulation*

LPS induces up-regulation of endogenous sialidases that cause desialylation and significantly enhance the secretion of pro-inflammatory cytokines such as TNF-α and IL-6 [350, 351]. In this study, we profiled the sialylation status of THP-1 macrophages upon LPS stimulation by quantifying total cellular Sias by LC-MS/MS and cell surface Sias by lectin staining and flow cytometry analysis. Further, we modulated the cellular sialylation status of macrophages at rest and LPS stimulation conditions by culturing the cells with free sialic acid (Neu5Ac and Neu5Gc) and sialidase inhibitors (Neu5Ac2en and Oseltamivir phosphate) to seek the sialylation effects on the cell elasticity and functions. Neu5Ac is a human sialic acid, while Neu5Gc is found in non-human mammalians, as humans lack Cytidine monophospho-N-acetylneuraminic acid hydroxylase (CAMH) [240]. Despite the inability to make Neu5Gc in humans, it has been identified in specific populations[9]. It was found that humans incorporate Neu5Gc into glycoconjugates via Sia metabolism pathway through diet, indicating a substrate tolerance of certain enzymes in the sialylation pathway [239]. Therefore, non-natural Sias can be used for metabolic modulation of cell surface Sia with designed structures [373]. First, we supplemented THP-1 macrophages at rest and under LPS-stimulation with Neu5Ac to modulate the Sia levels and elucidate the effect of sialylation on cell mechanic properties. Next, we explored the metabolic incorporation of Neu5Gc into the THP-1 macrophages to seek how altered sialylation will modulate cell biomechanics and functions. Results suggest that treatment with Neu5Ac modestly enhanced the Neu5Ac sialylation level of THP-1 macrophages at rest condition (Figure 3. 1A). LPS-stimulation dramatically reduced the Neu5Ac sialylation level of THP-1 macrophages. However, treatment with Neu5Ac rescued the Neu5Ac sialylation level of THP-1 macrophages stimulated with LPS, albeit to levels significantly lower than that in LPS absence.

On the other hand, treatment with Neu5Gc did not change the Neu5Ac sialylation level of THP-1 macrophages in both at rest and LPS-stimulation conditions. Interestingly, the Neu5Gc sialylation level of THP-1 macrophages increased dramatically, both in those at rest and LPS-stimulation conditions (Figure 3. 1B). Here too, LPS-stimulation led to a significantly lower Neu5Gc sialylation level of THP-1 macrophages compared to the at-rest condition (Figure 3. 1B). In general, Neu5Gc levels are at least two orders of magnitude lower than Neu5Ac levels at all conditions, which is to be expected as we measured these levels in human cells. However, there are two possible reasons for the lower Neu5Gc sialylation level in the macrophage. First, a lower intake or sialylation activity could cause lower Neu5Gc sialylation status. Second, it may be due to high sialidase activity induced by LPS stimulation. It is known that endogenous sialidase activity increases in cells of the immune system during cell activation [35, 102, 374]. Although oseltamivir is not an inhibitor for mammalian endogenous sialidase, it shows a certain level of inhibition of sialic levels and therefore often used in human cell cultures [35]. Neu5Ac2en is a general but less potent sialidase inhibitor for mammalian endogenous sialidase [102]. Therefore, we treated THP-1 macrophages with sialidase inhibitors Neu5Ac2en and oseltamivir phosphate to prevent desialylation from seeking the effect of sialylation on cellular mechanic properties and functions. As a result, neither Oseltamivir phosphate nor Neu5Ac2en showed inhibition on total Neu5Ac and Neu5Gc desialylation level of THP-1 macrophages under LPS stimulation (Figure 3. 1A and B).

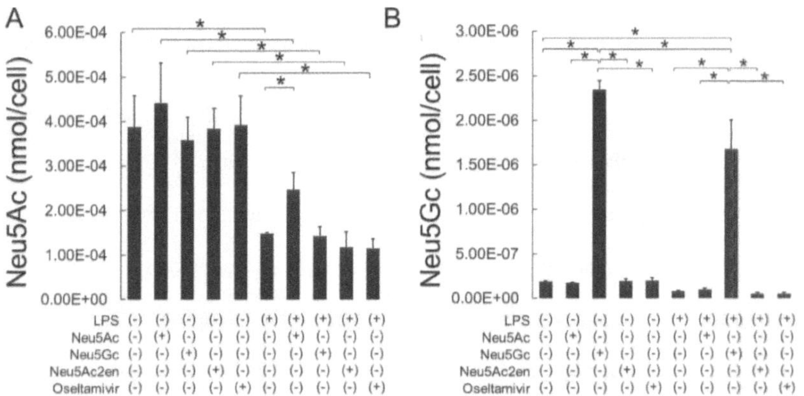

Figure 3. 1 Cellular sialylation levels of THP-1 macrophages upon LPS stimulation and metabolic modulation with free sialic acid (Neu5Ac, Neu5Gc) and sialidase inhibitors (Neu5AC2en, Oseltamivir phosphate). *Indicates $p < 0.05$ between the groups. Data were presented as average ± SE (n = 3 wells/ condition), with three independent repeats of the experiment.

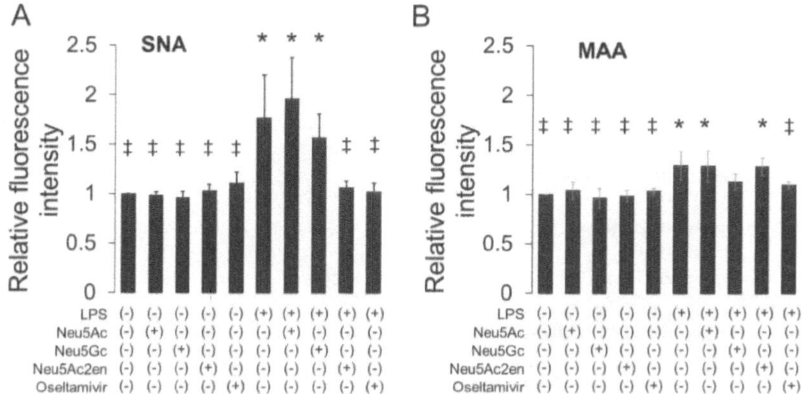

Figure 3. 2 Cell surface sialylation levels of THP-1 macrophages upon LPS stimulation and metabolic modulation with free sialic acids (Neu5Ac, Neu5Gc) and sialidase inhibitor (Neu5AC2en, Oseltamivir phosphate). *Indicates $p < 0.05$ vs. controls which received no exogenous supplements, while ‡ indicates $p < 0.05$ vs. LPS alone treated cells. Data were presented as average ± SE (n = 3 wells/ condition), with three independent repeats of the experiment.

The cell surface sialylation level changes of THP-1 macrophages upon LPS-stimulation and metabolic modulation were investigated by lectin staining and flow cytometry analysis. FITC-labeled Maackia Amurensis (MAA) and FITC-labeled Sambucus Nigra (SNA) lectin that specifically bind to α2,3-linkaged and α2,6-linkaged Sias, respectively, were used to investigate cell surface Sias. Results showed that α2,6-linkaged Sia dramatically increased upon LPS-stimulation (Figure 3. 2A). Neu5Ac alone or Neu5Gc alone had no effect on control cells, but modestly increased the levels of high α2,6-linkaged Sia of THP-1 macrophages when stimulated with LPS (Figure 3. 2A). Treatment with or without LPS showed no change of α2,6-linkaged Sia levels of THP-1 macrophages in the presence of both the sialidase inhibitors, as it may prevent desialylation (Figure 3. 2A). For α2,3-linkaged Sia level, broadly similar changes were observed, but the magnitude is much smaller compared to α2,6-linkaged Sia level (Figure 3. 3B). Overall, the cells remained viable after LPS treatment (100 ng/mL) over 3-day cultures with different Sias (500 μM).

3.3.2 Biophysical characteristics of THP-1 macrophages upon LPS stimulation and metabolic sialylation modulation

Representative force-indentation curves under various treatment conditions were shown in Figure 3. 3A. Elastic moduli (E_Y; Figure 3. 3B) of THP-1 macrophages upon LPS stimulation and metabolic modulation with free sialic acids (Neu5Ac and Neu5Gc) and sialidase inhibitors (Oseltamivir phosphate and Neu5Ac2en) was quantified from the force-indentation curves obtained from live single cell nano-indentation assays. The average modulus of as-derived macrophages significantly increased (1.63-fold) upon LPS-treatment. Compared to respective controls, adding Neu5Ac, Neu5Gc, or sialidase inhibitor to cell cultures significantly decreased elastic modulus of macrophages, both in the presence or absence of LPS ($p < 0.05$ vs. controls). The E_Y of non-LPS treated cells were significantly lower in all the cases compared to LPS-treated cells. No significant differences were noted between E_Y of Neu5Ac vs. Neu5Gc added cells, or Neu5Ac2en vs. oseltamivir treated cells, both in the presence and absence of LPS. Similar trends were noted in the forces of adhesion (F_{ad}, Figure 3. 4A), tether forces (F_T, Figure 3. 4B), and membrane tension (T_M, Figure 3. 4D) across the cases tested. Expectedly, opposite trends were noted for the radius of tether (as $R_T \propto 1/F_T$; Figure 3. 4C), i.e., the presence of Neu5Ac, Neu5Gc, or inhibitors significantly increased R_T of the tethers compared to controls, while the presence of LPS significantly lowered E_Y across all the cases. No significant differences in F_{ad}, F_T, R_T, and T_M were noted between Neu5Ac vs. Neu5Gc treated cells or Neu5Ac2en vs. oseltamivir treated cells.

Taken together, these results suggest that (i) cells became stiffer in the LPS-stimulated inflammatory environment, despite providing metabolic sialylation or sialylation inhibitors; (ii) metabolic sialylation or sialylation inhibitors significantly suppressed macrophage membrane characteristics (adhesion forces, tether forces, membrane tension) and cytoskeletal stiffness, even under inflammatory conditions, although such levels were higher in LPS presence; and (iii) no significant differences between the addition of Neu5Ac or Neu5Gc was evident.

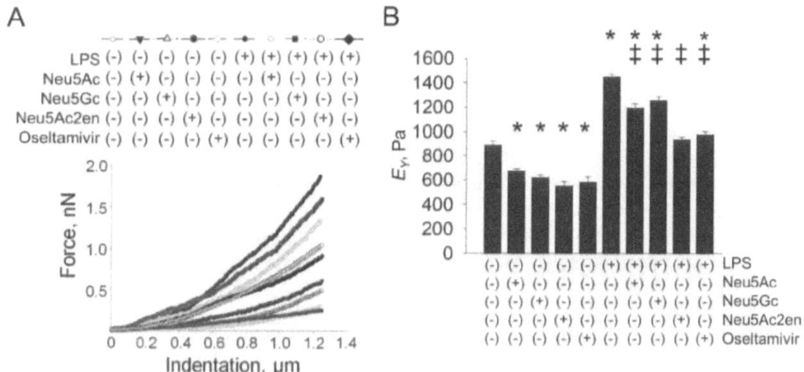

Figure 3. 3 (A) Representative force-indentation curves obtained at random locations on THP-1 macrophages upon LPS stimulation and metabolic modulation with free sialic acids (Neu5Ac, Neu5Gc) and sialidase inhibitor (Neu5AC2en, Oseltamivir phosphate). (B) Average ± SE of Young's modulus values calculated from such force curves (at least n = 40 cells/ condition). *$p < 0.05$ vs controls which received no exogenous supplements; ‡ $p < 0.05$ vs. LPS alone treated cells.

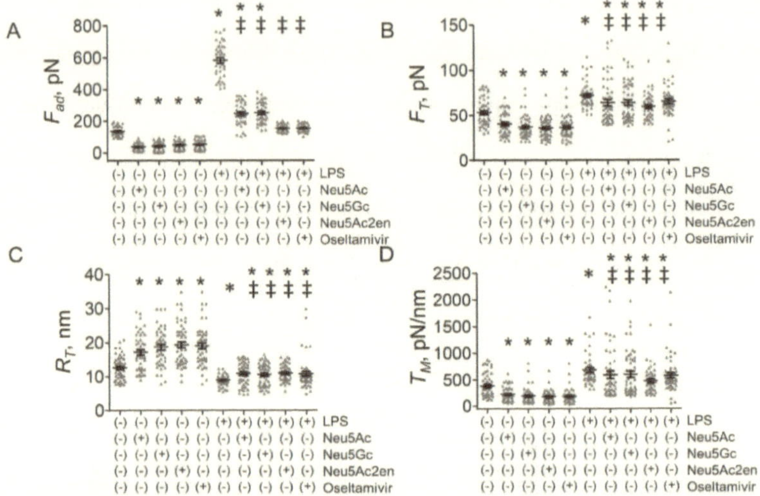

Figure 3. 4 Average ± SE of adhesive forces (F_{ad}, **A**), tether forces (F_T, **B**), the radius of tether (R_T, **C**), and membrane tension (T_M, **D**) of THP-1 macrophages upon LPS stimulation and metabolic modulation with free sialic acids (Neu5Ac, Neu5Gc) and sialidase inhibitor (Neu5AC2en, Oseltamivir phosphate) (at least n = 40 cells/ condition). * p < 0.05 vs controls which received no exogenous supplements; ‡ p < 0.05 vs. LPS alone treated cells.

Our observations agree with previous studies that reported the average elastic moduli of various macrophage cell lines in the 1.03 - 2 kPa range [34-36]. LPS-stimulation increased elastic modulus by two to three-fold [359]. Pi et al. observed both the adhesion forces and the elastic modulus of RAW264.7 macrophages to increase ~3-fold with 100 ng/mL LPS-stimulation [375]. However, the cells were fixed with paraformaldehyde before AFM testing. A 2.3-fold increase in adhesion force was reported [376] in human macrophages treated with 10 µg/mL LPS (AFM studies), similar to that noted in our study (~ 3.2-fold). Overall, our results highlight the significant changes induced by LPS treatment to macrophages and the role of Sias in modulating cell surface adhesive forces, tether forces, and membrane tension of macrophages under normal and inflammatory conditions.

3.3.3 *Cellular OVA antigen uptake of THP-1 macrophages upon LPS stimulation and metabolic modulation*

Cellular uptake of antigen is an important function of macrophages as antigen-presenting cells in adaptive immune responses. THP-1 macrophage activity was tested by cellular uptake of fluorescent model antigen (FITC-OVA) upon LPS stimulation and metabolic modulation [377]. Thus, changes in macrophage membrane elasticity might affect antigen uptake activity. It is known that LPS stimulation increases macrophage cell elasticity. In this study, a modest increase in the antigen uptake activity was noted with LPS-alone stimulation compared to controls. No significant changes in antigen uptake were noted when cells were treated with Sias, with or without the LPS-stimulation. Except for a marginal decrease in activity in the presence of Neu5Ac2en and LPS, exposure to oseltamivir phosphate had no effect on antigen uptake activity of THP-1 cells, both in the presence or absence of LPS.

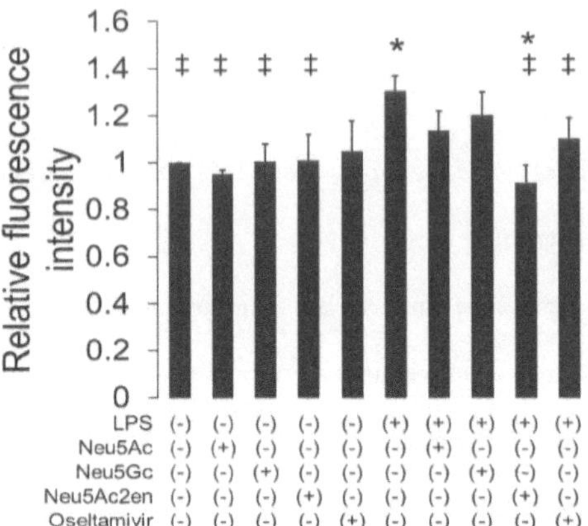

Figure 3. 5 Relative antigen uptake activity (FITC-OVA) of THP-1 macrophages (normalized to cultures that received no supplements) upon LPS stimulation and metabolic modulation with free sialic acid (Neu5Ac, Neu5Gc) and sialidase inhibitors (Neu5AC2en, Oseltamivir phosphate). Data were presented as average ± SE (n = 3 wells/ condition), with three independent repeats of the experiment. * p < 0.05 vs controls which received no exogenous supplements; ‡ p < 0.05 vs. LPS-alone treated cells.

3.4 Conclusion

In this study, we characterized the sialylation status of LPS-stimulated THP-1 macrophages and their elastic properties and function as well. Specifically, we confirmed that sialylation status is closely related to macrophage elastic properties and thus directly involved in macrophage functions. Further, we modulated macrophage sialylation status by feeding the cells with free sialic acid (Neu5Ac, Neu5Gc) and sialidase inhibitors and examined the impact on cellular elasticity and phagocytosis. Key observations from this study are as follows: results are taken together suggest that although THP-1 macrophage stimulation with 100 ng/mL LPS did not affect their viability, it nevertheless induced a significant reduction in Sia levels (Neu5Ac & Neu5Gc; Figure 3. 1) and elevation in α2,6-linkaged Sia expressions (Figure 3. 2), which positively contributed to robust increases in cell elastic modulus (Figure 3. 3), adhesion/tether forces and membrane tension (Figure 3.

4), which in turn contributed to relatively higher phagocytosis activity. Meanwhile, exogenous supplementation of Sias or Sia inhibitors to LPS-stimulated cells significantly and consistently reduced the measured biomechanical characteristics of these cells. Studies such as this unravel the mechanisms by which Sias modulate macrophage functions and establish the relationship between the biological, biomechanical, and functional outcomes of these cells when Sia levels are altered on their surfaces. Overall, a systematic recognition of sialylation status related to the cellular elasticity of macrophages will contribute to defining macrophage phenotypes and understand macrophage functional diversity. Further, it will provide novel mechanisms and approaches for disease diagnosis and treatment.

CHAPTER IV

METABOLIC SIALO-ENGINEERING OF MACROPHAGES AND THEIR PROPERTY AND FUNCTION

4.1 Introduction

Sialic acids (Sias) are a family of 9-carbon containing acidic monosaccharides and often terminate cell surface glycans of either glycoproteins or glycolipids [293]. Sias are critical components of glycoconjugates that serve as receptors involved in many biological processes. In addition, Sias can contribute to the stability of glycoconjugates, such as hormones or antibodies *in vivo*. Over 50 different naturally occurring members of Sias have been identified [378]. The C-5-amino derivative represents the well-known neuraminic acid, and its amino functional group can be either acetylated (*N*-acetylneuraminic acid, Neu5Ac) or glycosylated (*N*-glycolylneuraminic acid, Neu5Gc). Notably, humans make Neu5Ac but are incapable of synthesizing Neu5Gc [379]. However, certain human population incorporates Neu5Gc into glycoconjugates *via* the Sia metabolism pathway through dietary acquisition, indicating a substrate tolerance of certain enzymes in the sialylation pathway[239]. It has been suggested that the replacement of Neu5Ac with Neu5Gc residues in humans may profoundly alter biological processes as some molecules recognize glycoconjugates containing Neu5Gc and Neu5Ac with different affinities [349].

Metabolic sialo-engineering has been developed for modifying cell surface Sia with non-natural Sias [380]. This provides a very useful tool to investigate the biological function of Sia as part of sialoglycoconjugates in living cells. Metabolic modification of the *N*-acyl group of Sias can be achieved through the administration of Sia precursor analogues *N*-modified D-mannosamines, which are taken up, metabolized, and incorporated into cell surface sialoglycoconjugates of mammalian cells. In addition, Sias carrying modifications in the C-5 position can also be metabolically incorporated onto cell surface sialoglycoconjugates. One key advantage of using Sia analogues over Sia precursor analogues for the metabolic engineering of cellular Sias is their independence from the multistep biosynthetic pathway. A range of non-natural Sia analogs has been successfully installed in a variety of cells and organisms [381].

Macrophages play a pivotal role in inflammation, tissue injury, repair processes, and the immune system [355]. Macrophage cell surface expresses a dense layer of glycans often terminated with Sias, which can mediate many biological processes of macrophages, such as host-pathogen recognition, migration, and antigen presentation, among other non-immune related processes [378]. Recently, we confirmed that cell surface sialylation status is closely related to macrophage biomechanical characteristics (elastic modulus, tether force, tether radius, adhesion force, and membrane tension) and thus directly involved in macrophage function [326]. Therefore, structural variations of cell surface Sias of macrophages can profoundly affect the biological function of the glycoconjugates and the specific function of macrophages as well.

In this study, we synthesized a Sia derivative with *N*-Glycine at the 5-position (Neu5Gly) and investigated its metabolic incorporation together with human Sia Neu5Ac and non-human Sia Neu5Gc into human THP-1 macrophage cell lines, which is widely used as the macrophage cell model as it exhibits immune properties similar to native monocyte-derived macrophages [369](Figure 4. 1). The primary amine at 5-position of Neu5Gly together with the 1-carboxylic acid makes the molecule zwitterionic. It is well known that zwitterionic carbohydrates show potential immune activities. For example, zwitterionic polysaccharides (ZPS), such as the capsular polysaccharides of many strains of Bacteroides fragilis, Staphylococcus aureus, and Streptococcus pneumoniae type 1, elicit potent CD4(+) T cell responses *in vivo* and *in vitro* [382]. The cell-mediated response to ZPS depends on the presence of positively charged and negatively charged groups on each repeating unit of the polysaccharide [382]. Therefore, it is expected that this ziwitterionic Neu5Gly, if taken by the cells, will affect the biological function of the glycoconjugates and the cellular function of macrophages. The ability to modulate cell surface sialylation with zwitterionic Sia will be a useful tool to modify the cell surface Sia function of macrophages.

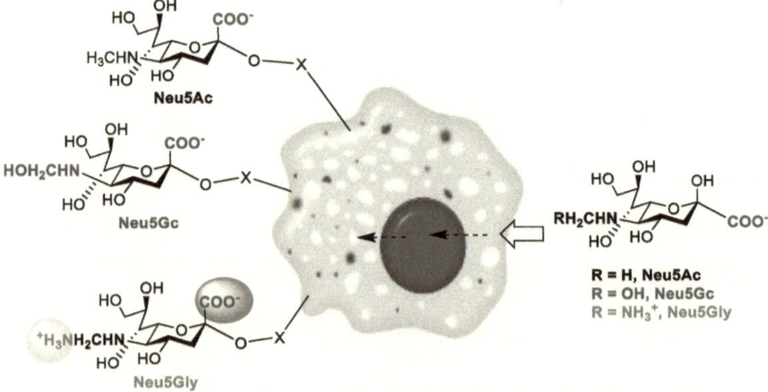

Figure 4. 1 Metabolic modification of cell surface Sias of THP-1 macrophages.

4.2 Experiments

4.2.1 *Materials and reagents*

FITC-labeled *maackia amurensis* (MAA) lectin was purchased from bioWORLD (Dublin, OH, USA). FITC-labeled *sambucus nigra* (SNA) lectin was provided by Vector Laboratories (Burlingame, CA, USA). N-Acetyl-D-neuraminic acid (Sia, 98%) was supplied by Rose Scientific Ltd (Edmonton AB, Canada). N-Acetyl-D-neuraminic acid-1,2,3-$^{13}C_3$ ($^{13}C_3$-Sia, 99%), 4′,6-diamidino2-phenylindole (DAPI), Methanol (HPLC grade), paraformaldehyde (PFA, 16% w/v), Fluorescein Phalloidin, 4′,6-Diamidino-2-Phenylindole, Fluorescein conjugate Escherichia coli (K-12 strain) BioParticles™ and acetic acid (ACS grade) were obtained from Fisher Scientific (Hanover Park, IL, USA). Ketodeoxynonulosonic acid (KDN, 99%), 3,4-diaminotoluene (DAT, 97%), 4- bromo-1,2-diaminobenzene (97%), 4-nitro-O-phenylenediamine (98%), 2-mercaptoethanol (99%), Dihydrochloride (DAPI), Phorbol 12-myristate 13-acetate (PMA), Triton X-100 and sodium hydrosulfite (85%) were purchased from Sigma–Aldrich Inc. (St. Louis, MO, USA). Deionized water was generated from Barnstead Nano-PURE Water Purification System (Asheville, NC, USA). Phosphate-buffered saline (PBS) was obtained from Cleveland Clinic (Cleveland, OH, USA). Bovine Serum Albumin (BSA) was purchase from Santa Cruz Biotechnology Inc. (Dallas, Texas, USA).

4.2.2 *Apparatus*

Flow cytometry was performed on a FACSCanto II system consisting of a blue and a red laser, operated through BD FACSDiva software (BD Bioscience, Mountain View, CA, USA). Images were acquired at a 60× magnification using a Nikon A1Rsi confocal microscope, and data analysis was analyzed with NIS-Elements software (Nikon

Instruments Inc., Melville, NY, USA). LC-MS/MS quantification was carried out by a Nexera liquid chromatography system (Shimadzu, Columbia, MD, USA) interfaced to a Qtrap 5500 mass spectrometer equipped with an electrospray ionization source (AB SCIEX, Framingham, MA, USA), and the data acquisition and processing were conducted using Analysis 1.6.1 software. HPLC-fluorescence qualification was performed on the Shimadzu LC20 system (Shimadzu UAS MFG Inc., Canby, USA), which consisted of a CBM-20A controller, two LC-20AD pumps, a SIL-20AC HT autosampler, and the RF-10AxL detector. MTT assay was performed on SpectraMax® Plus 384 Microplate Reader (Walpole, MA, USA), and data were analyzed on SoftMax Pro 6.3.

4.2.3 *THP-1 monocytes differentiation*

1×10^6 THP-1 monocytes were seeded into each well into the six wells plates (for MTT assay, flow cytometry, HPLC, and LC-MS/MS), or with a cover slide (for Confocal microscopy) and treated with 10nM PMA for 48 hours to differentiate to THP-1 macrophage cells as described in our previous report.[9]

Sias Incorporation of THP-1 macrophage: THP-1 macrophage cells were treated with 0, 20, 50, 100, or 500 µM of the different Sias for 72 hours.

4.2.4 *MTT assay*

To study the toxicity of Neu5Ac, Neu5Gc, and Neu5Gly, THP-1 monocytes and the differentiated macrophages were treated with each compound at concentrations of 20, 50, 100, 250, and 500 µM. After incubation for one, two, or three days, cell viability was tested with MTT assay following the manufactory's instruction. Briefly, cells were incubated with 0.5 mg/ml MTT for 4 h at 37 °C. The resulting formazan was solubilized with 100% DMSO, and the concentration was determined by optical density at 570 nm.

4.2.5 *Membrane isolation*

The Crude membrane fraction was pellet by centrifugation 10,000 ×g for 15min, and the pellet was washed with 200 μL DI water three times, and the pellet was re-suspended into 200μL of DI water.

4.2.6 *Sias release and derivatization*

Briefly, Neu5Ac, Neu5Gc, or Neu5Gly was released from cell lysates by hydrolyzing with 2 M acetic acid (1:1) at 80°C for 90 min and then derivatized with 3,4-diaminotoluene (DAT) at 80 °C for 40 min. An aliquot of 5μl of the mixture was subjected to LC-MS/MS quantification, or 10μl of the mixture was an injection.

4.2.7 *HPLC condition*

The separation was carried out using an XSelect HSS T3 column (3.5 mm particle size, 2.1 mm × 100 mm, Waters Corporation, Milford, USA) with an XSelect HSS T3 Guard Column (3.5 mm particle size, 2.1 mm × 10 mm, Waters Corporation, Milford, USA). The mobile phase was a gradient mixture of phase A (deionized water containing 80 mM NH_4HCO_3) and Phase B (methanol) with a flow rate of 0.2 mL/min. The gradient time program was given in Table 4. 1. The fluorescence detection was set at 379 nm for excitation and 432 nm for emission wavelength for Sia-DAT detection.

4.2.8 *LC-MS/MS condition*

Due to Neu5Ac in nature, $^{13}C_3$-Neu5Ac was used as a surrogate standard to generate the calibration curve for Neu5Ac quantification. A similar structure compound, ketodeoxynonulosonic acid (KDN), was used as the internal standard (IS) for all the quantifications. The DAT derivatized $^{13}C_3$-Neu5Ac, KDN, Neu5Ac, Neu5Gc, and Neu5Gly were separated by a Phenomenex Kinetex C18 column (2.1 × 50 mm, 1.7 μm) with a binary linear gradient elution, in which phase A was deionized water containing 10 mM ammonium formate and phase B was methanol. The gradient program (Table 4. 2.) started at 10% B and was maintained for 1 min, and then changed linearly to 95% B in 3 min and maintained for 2 min. In order to equilibrate the column, the gradient sharply dropped to 10% B in 0.1 min and kept for 2.9 min. The total run time was 8.9 min for each sample. The MS detection was carried out in positive electrospray ionization and multiple reaction monitoring (MRM) mode. The quantitative MRM transitions were set at m/z 399.1 $\rightarrow$ 256.1 for $^{13}C_3$-Neu5Ac, 355.1 $\rightarrow$ 175.0 for KDN, 396.1 $\rightarrow$ 253.0 for Neu5Ac, 412.2 $\rightarrow$ 253.1 for Neu5Gc, and 411.2 $\rightarrow$ 283.1 for Neu5Gly.

Due to the structural similarity of Neu5Gc and Neu5Gly, we were interested to see the co-incubation effect of these two compounds. THP-1 monocytes and the differentiated macrophages were treated with both Neu5Gc and Neu5Gly at a concentration of 500 μM for each compound. As controls, the cells treated with vehicle (PBS), Neu5Gc, or Neu5Gly only at the same concentration. After three days of incubation, Neu5Ac, Neu5Gc, and Neu5Gly in each sample were analyzed by LC-MS/MS.

Table 4. 1 HPLC gradient time program for HPLC-Fluorescence.

Time (min)	Methanol concentration (%)
0	5
5	30
10	35
20	35
25	60
30	98
35	98
35.01	5
40	5

Table 4. 2 HPLC gradient time program for LC-MS/MS.

Time (min)	Methanol concentration (%)
1	10
4	100
6	100
6.1	10
9	stop

4.2.9 *Lectin binding assay*

THP-1 macrophage cells were incubated with (20µg/mL) FITC labeled MAA or SNA lectins for 15 minutes at room temperature, followed by PBS washing at least twice.

Phagocytosis Assay: The macrophage phagocytosis was tested following the manufactory's instruction. Briefly, cells were incubated with FITC labeled *E. Coli* for 1.5 hours at 37 °C, with PBS washing at least two times.

4.2.10 *Flow cytometry*

THP-1 macrophage cells were gently removed from the plate by incubated in the 1.5 mL cold PBS and blow off by pipette. The THP-1 macrophage cells were fixed with 3.7% formaldehyde, washed with PBS at least twice. Then THP-1 macrophage was treated with 0.1% Triton X-100 in PBS for 3-5 minutes, washed with PBS three times. THP-1 macrophage cells were re-suspended into 1.5 mL of PBS contains (0.5% BSA and 0.5% NaN3) and separated into three different tubes.

4.2.11 *Confocal microscopy*

THP-1 macrophage cells were fixed with 3.7% formaldehyde, washed with PBS at least two times. The cover slide was treated with 0.1% Triton X-100 in PBS for 3-5 minutes, washed with PBS three times. The Fluorescent phallotoxins stock solution was 1:40 dilute into the 1% BSA in PBS. The dilute solution (200 µL) was placed on each slide, and the slide was incubated for 20 minutes at room temperature, followed by PBS washing three times. The cover slide was treated with DAPI stock solution 1:50 dilution into PBS, washed with PBS three times. Finally, the cover slides were put on the slide and observed under the Confocal Microscope.

4.2.12 *Statistical analysis*

Unless otherwise noted, all data was represented as average standard error (SE) from n=3 wells/conditions, with at least three independent repeats of each assay. Significance in differences between groups was analyzed using two-way analysis of variance (ANOVA) with Tukey post-hoc HSD test, and p-values < 0.05 were considered statistically significant.

4.3 Results and Discussion

To achieve the desired placement of the glycine on the C-5 position, the removal of the C-5 acetamide of Neu5NAc was required (Scheme 1). Initially, a literature method [383] was followed, in which Barium hydroxide ($Ba(OH)_2$) in methanol-water (1:1, v/v, reflux) was used but did not give the desired benzyl α-glycoside of neuraminic acid **3** instead of a mixture of unidentified compounds. We also checked with MsOH in methanol but did not get the desired benzyl α-glycoside of neuraminic acid **3** either. Fortunately, reaction with Boc_2O and DMAP in THF went smoothly, followed by deacetylation with sodium methoxide in methanol affording *N*-Boc-protected benzyl α-glycoside of neuraminic acid **4**. The reaction of benzyl α-glycoside of neuraminic acid **4** with the 4-nitrophenyl ester of N-benzyloxycarbonyl glycine followed by catalytic hydrogenation produced Neu5Gly in 41.4% overall yield (Scheme 1). All compounds were characterized by ^1H and ^{13}C NMR spectra.

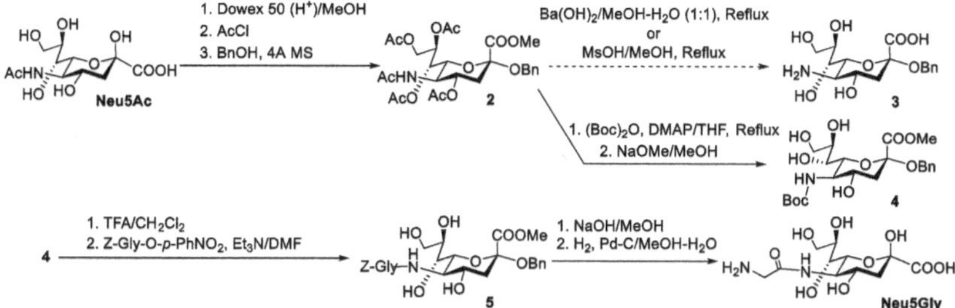

Scheme 4. 1 Synthesis of Neu5Gly, Sia analogue with a N-glycine at the 5-position.

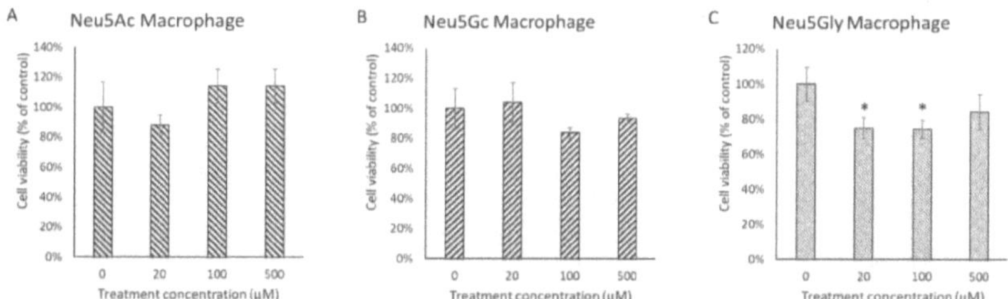

Figure 4. 2 MTT assay of THP-1 macrophages treated with different concentrations of Sias: A. Neu5Ac; B. Neu5Gc; C. Neu5Gly. *Indicates $p < 0.05$ between Sia treatment groups to the non-treatment group. Data points represent the mean ± SD, n=3.

Before the cell incorporation study, we examined the cytotoxicity of Neu5Gly together with Neu5Ac and Neu5G to the THP-1 macrophage cells at various concentrations. As a result, less toxicity was observed for the cells treated with Neu5Ac, Neu5Gc, and Neu5Gly up to 500 µM in the MTT assay (Figure 4. 2).

The incorporation of Neu5Ac, Neu5Gc, and Neu5Gly in the THP-1 macrophage cells was examined by our newly developed Sia derivatization and HPLC analysis[323]. In this study, the THP-1 macrophage cells were cultivated in the presence of Neu5Ac, Neu5Gc, and Neu5Gly (500 mM), respectively. After 72 h, the cells were extensively washed, and the cells were lysate, and free Sias were released from the cell lysate by hydrolyzing with

2 M acetic acid (1:1) at 80 °C for 90 min, and then derivatized with 3,4-diaminotoluene (DAT) at 80 °C for 40 min, and were subjected for HPLC analysis. Both Neu5Ac, Neu5Gc, and Neu5Gly were eluted as discrete peaks (Figure 4. 3). The peaks coincided with authentic Neu5Ac, Neu5Gc, and Neu5Gly standards (Figure 4. 3 A) and cell lysate samples (Figure 4. 3C), respectively. This result indicates that uptake of Neu5Gc and Neu5Gly could be incorporated and resulted in the structural modification of Sias on the cell surface. We examined whether contamination by input Sia analogues during the purification procedure could account for these results. Neu5Ac, Neu5Gc, and Neu5Gly were added to the THP-1 macrophage cells but without incubation (0 h), which were immediately followed by cell harvesting and cell lysis, and HPLC profiling. Except for the original Neu5Ac, Neither Neu5Gc nor Neu5Gly was detectable in the cell lysates (Figure 4. 3B). However, after 72 h incubation, Neu5Gc and Neu5Gly were detected in the cell lysates (Figure 4. 3C). These results validated our experimental procedures and confirmed that Sias were metabolically incorporated into cellular glycoconjugates.

As a direct and definitive way of assessing the incorporation of these Sias into cell surface glycoconjugates, we analyzed the membrane-associated Sia fraction of cells by the same Sias derivatization and HPLC analysis. After 72 h incubation, the cells were extensively washed, and the cell membrane fractions were isolated, and free Sias was released from the cell membrane by hydrolyzing with 2 M acetic acid (1:1) at 80 °C for 90 min, and then derivatized with 3,4-diaminotoluene (DAT) at 80 °C for 40 min, and were subjected for HPLC analysis. As a result, Neu5Ac, Neu5Gc, and Neu5Gly were eluted as discrete peaks in the cell membrane samples (Figure 4. 3D), respectively. This result indicates that uptake of Neu5Gc and Neu5Gly could be incorporated and resulted in the structural modification of Sias on the cell surface. Unfortunately, HPLC data could not be used for quantitative analysis due to the low incorporation of Neu5Gly.

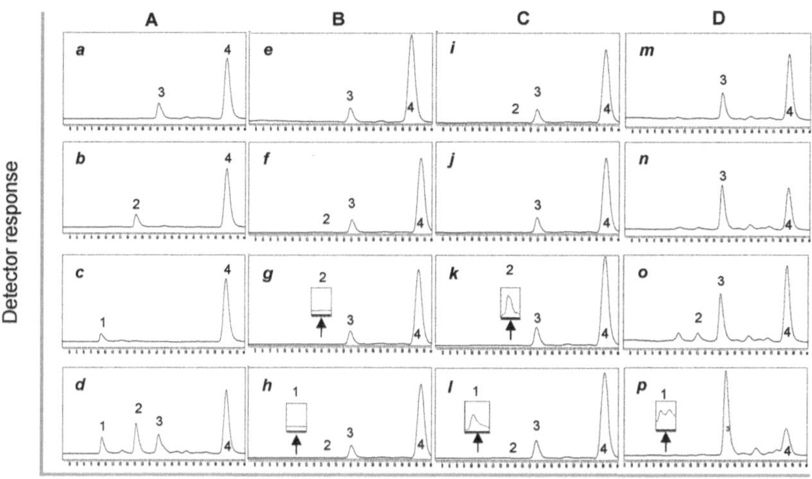

Retention time (min)

Figure 4. 3 Sia incorporation in cell surface and cell lysate glycoconjugates of THP-1 macrophages. Representative HPLC chromatogram showing separation of Neu5Ac, Neu5Gc, and Neu5Gly derivatized with DAT at 80 °C for 1.5 h. (**A**) <u>**Standards**</u>: Neu5Ac (**a**) Neu5Gc (**b**), Neu5Gly (**c**) and the mixture of Neu5Ac, Neu5Gc, and Neu5Gly (**d**); (**B**) <u>**Cell lysates with 0 h Sias treatment**</u>: PBS treated THP-1 macrophages (**e**), Neu5Ac treated THP-1 macrophages (**f**), Neu5Gc treated THP-1 macrophages (**g**), and Neu5Gly treated THP-1 macrophage (**h**); (**C**) <u>**Cell lysates with 72 h Sias treatment**</u>: PBS treated THP-1 macrophages (**i**), Neu5Ac treated THP-1 macrophages (**j**), Neu5Gc treated THP-1 macrophages (**k**), and Neu5Gly treated THP-1 macrophages (**l**); (**D**) <u>**Cell membrane samples with 72 h Sias treatment**</u>: PBS treated THP-1 macrophages (**m**), Neu5Ac treated THP-1 macrophages (**n**), Neu5Gc treated THP-1 macrophages (**o**), and Neu5Gly treated THP-1 macrophage (**p**). Peaks: 1 from Neu5Gly-DAT, 2 from Neu5Gc-DAT, 3 from Neu5Ac-DAT, 4 from DAT.

Next, the incorporation of Neu5Ac, Neu5Gc, and Neu5Gly into THP-1 macrophages was quantitatively investigated with LC-MS by treating THP-1 macrophages with Neu5Ac, Neu5Gc, and Neu5Gly at the concentration of 0, 20, 50, 100, 250, and 500 μM for three days, respectively. The concentration of Neu5Ac, Neu5Gc, and Neu5Gly in the treated

79

cells was determined by our previously established LC-MS/MS method with a few modifications [101]. Briefly, free Neu5Ac, Neu5Gc, and Neu5Gly were released from cell lysates by hydrolyzing with 2 M acetic acid (1:1) at 80 °C for 90 min and then derivatized with 3,4-diaminotoluene (DAT) at 80 °C for 40 min. An aliquot of 5 µL of the mixture was subjected to LC-MS/MS quantification. As a result, there was no significant change when cells were treated with Neu5Ac (Figure 4. 4A), while the Neu5Gc level increased dramatically with the increase of Neu5Gc concentration (Figure 4. 4B). The Neu5Ac level is insensitive to change in metabolic flux because the Golgi concentration of CMP-Neu5Ac exceeds the K_m for sialyltransferases [373, 384]. As a consequence, enhanced CMP-Neu5Ac level does not translate into increased Neu5Ac incorporation. However, the human cell is in the absence of CMAH. The CMP-Neu5Gc was newly made under CMAS's substrate tolerance range, which rallied the incorporation of Neu5Gc. This concentration-dependent incorporation of Neu5Gc of human macrophage cell lines further supports the human intake of this non-human Sia. Interestingly, concentration-dependent Neu5Gly incorporation was also found for THP-1 macrophages as determined by LC-MS/MS method (Figure 4. 4C). The incorporation scale was lower than Neu5Gc as a comparison.

Further, we investigated whether Neu5Gly as a Neu5Gc analog would metabolically inhibit the Neu5Gc incorporation into THP-1 macrophages. As a result, the concentration of Neu5Gc in cells treated with Neu5Gly unchanged as compared with non-treated, indicating no inhibition of the Neu5Gc incorporation THP-1 macrophages. Also, we found that the Neu5Ac treatment did not affect the Neu5Gc incorporation as the ratio of Neu5Ac/Neu5Gc did not change even with the increase of Neu5Ac concentration. (Table 4. 3) Similarly, Neu5Gly treatment did not change the ratio of Neu5Ac/Neu5Gc, indicating that there is no competitive inhibition between Neu5Gc and Neu5Gly in their incorporation

(Table 4. 3).

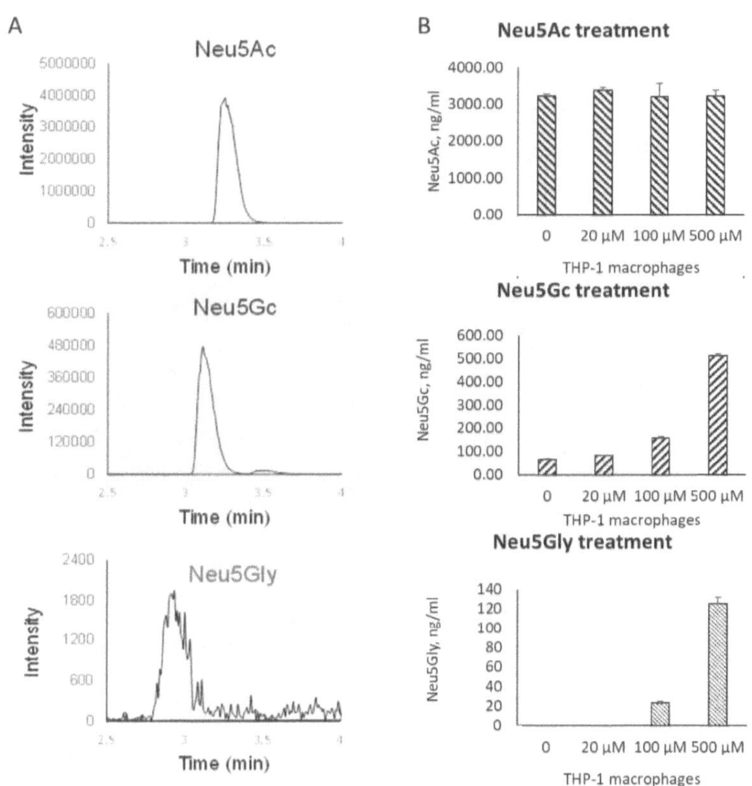

Figure 4. 4 (A) Representative LC-MS/MS chromatogram showing DAT derivative of Neu5Ac, Neu5Gc, and Neu5Gly in cell lysis samples of THP-1 macrophages treated with Neu5Ac, Neu5Gc, and Neu5Gly, respectively. **(B)** Sia incorporation of THP-1 macrophages analyzed by LC-MS/MS: THP-1 macrophages treated with different concentrations of Neu5Ac, Neu5Gc, and Neu5Gly. Data points represent the mean ± SD, n=3.

Table 4. 3 Sias of THP-1 macrophages upon the treatment with Sias.

Ratio of Cell Surface Sias	Treatment with Sia			
	0.	20	100	500 µM
Neu5Ac/Neu5Gc (treat with Neu5Ac)	43	43	45	45
Neu5Ac/Neu5Gc (treat with Neu5Gc)	41	32	18	5
Neu5Ac/Neu5Gc (treat with Neu5Gly)	41	41	41	42
Neu5Ac/Neu5Gc/Neu5Gly (treat with Neu5Gly)	-	-	41/1/0.3	45/1/1.6

[a] analyzed by LC-MS/MS

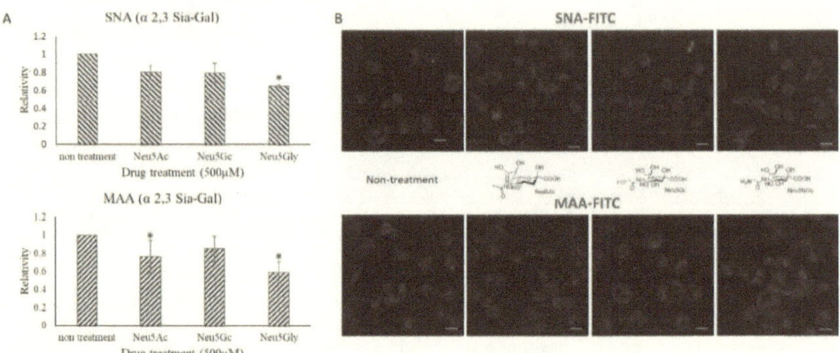

Figure 4. 5 Lectin binding of THP-1 macrophages upon different Sias treatment. (**A**) Flow cytometry analysis for SNA binding to cell surface α2,6Sia-Gal of the cells and MAA binding to cell surface α2,3Sia-Gal of the cells treated with 500 µM of different Sias; (**B**) confocal microscopy image for the SNA (blue) binding to cell surface α2,6Sia-Gal of the cells and the MAA (blue) binding to cell surface α2,3Sia-Gal of the cells treated with 500 µM of different Sias. The nucleus is stained with DAPI (blue). *Indicates $p < 0.05$ between Sia treatment groups to the non-treatment group. Data points represent the mean ± SD, n=3. (bar size 20µm)

As the terminal Sia structure change confirmed, we next analyzed the lectin binding capacity upon cell surface Neu5Gc and Neu5Gly incorporation. Lectins are proteins that bind to the specific glycans in a highly selective fashion. SNA lectin recognizes α2,6-linked Sia to galactose, while MAA lectin selectively binds to α2,3-linked Sia. Both the flow cytometry and confocal data showed that the lectin binding capacity changed upon cell surface Neu5Gc and Neu5Gly incorporation (Figure 4. 5). All the Neu5Ac, Neu5Gc, and Neu5Gly treatments caused reduced SNA and MAA lectin bindings. Among them, Neu5Gly treatment resulted in the lowest lectin bindings. It is unclear why the Neu5Ac treated cells showed reduced lectin binding even there were neither Neu5Ac nor Neu5Gc changes upon Neu5Ac treatment. It is clear that the Neu5Gc and Neu5Gly incorporation showed less lectin binding.

Macrophage has an essential role in phagocytosis, and we analyzed whether the cell surface sialylation status change could affect its phagocytosis capacity. Commercially available FITC-labeled *E. coli* was used in this study. After 24 h treatment of THP-1 macrophages having different Sias with FITC-labeled *E. coli*, we analyzed their phagocytosis by THP-1 macrophages by flow cytometry analysis and confocal image. As a result, THP-1 macrophages treated with different Sias showed increased FITC-labeled *E. coli* intakes as compared to LPS treated. However, there was no significant difference found for each Sia treatment (Figure 4. 6). This may be due to the expression level of Neu5Gc and Neu5Gly are still quite lower than the main Sia Neu5Ac. A high yield of incorporation of Neu5Gc and Neu5Gly may significantly affect macrophages functions, including phagocytosis capacity, which deserves future research.

4.4 Conclusion

In this study, we systematically examined metabolic modification of cell surface sialoform of THP-1 macrophages with human and non-human Sia, Neu5Ac, Neu5Gc, and Neu5Gly, respectively. HPLC analysis identified the Sias incorporation in membrane glycoproteins and glycolipids. LC-MS/MS quantification confirmed the total incorporation of Sias. We found that human THP-1 macrophage cell lines intake Neu5Gc and Neu5Gly in a concentration-dependent way. These cell surface sialoform changes affect their specific lectin binding confirmed under flow cytometry and confocal imaging analysis. Also, the phagocytosis capability changed as compared to non-treated cells. Although the incorporation yield is low for THP-1 macrophages, metabolic incorporation of zwitterionic Sia for other cells need to be investigated as it provides a new chemical functionality of glycan chain on the cell surface and new functionality of the cells as well.

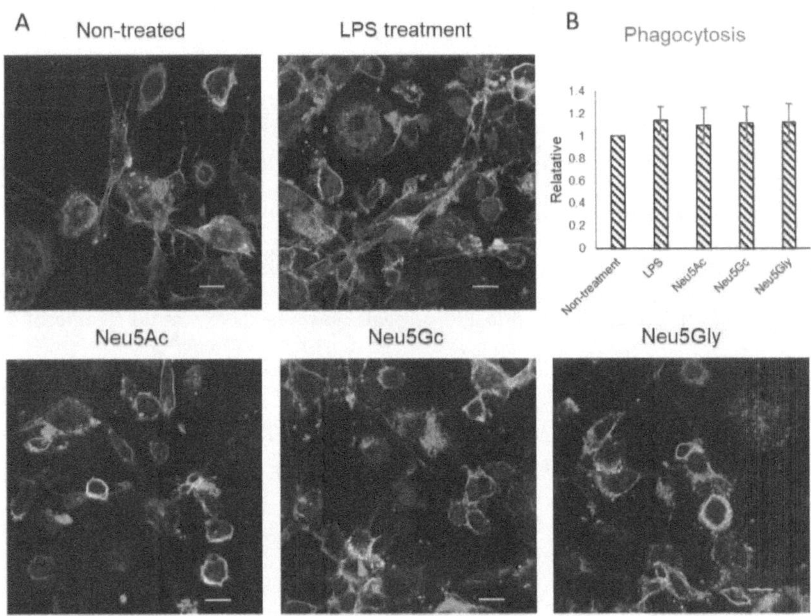

Figure 4. 6 Macrophage phagocytosis after treated with the 500µM of different Sias. (A) Confocal image of HP-1 macrophage phagocytosis of FITC-labeled E. coli upon different Sias treatment; (B) Flow cytometry analysis of THP-1 macrophage phagocytosis of FITC-labeled *E. coli* (green) upon different Sias treatment. The skeleton of THP-1 macrophages is stained with FITC-phallotoxins (yellow). The nucleus is stained with DAPI (blue). Data points represent the mean ± SD, n = 3. (bar size 20µm)

CHAPTER IV

SUMMARY

Sias was often found as the terminal sugars of the glycan structure of the glycoconjugates in different linkage and levels, known as sialoform. The sialoform of the glycoconjugates commonly controlled by sialylation and desialylation processes. Sias play essential roles in the physiological and pathological process, including immunological process and tumor progression. Sias are highly involved in the innate immune response by interacting with the Siglec in monocytes and macrophages. Meanwhile, the removal of Sias from the glycans catalyzed by sialidase (desialylation) is also involved in many biological processes through changing protein conformation and assembling and unknown mechanisms. In mammalian cells, there are four kinds of sialidase, which are Neu1, Neu2, Neu3, and Neu4. Neu1 is the lysosomal sialidase, which degrades the sialoglycoconjugates in the lysosome; Neu2 is the cytosolic sialidase that is involved in desialylation of sialoglycoconjugates in the cytosol; the membrane-associated Neu3 is the sialidase located at the membrane and specific for gangliosides; Neu4 is found on the outer membrane of mitochondria, which could work on a bunch of sialoglycoconjugates including glycoproteins, gangliosides, and oligosaccharides. In addition to specific cellular location, endogenous lysosomal sialidase Neu1 could move to the cell suface upon cell activation in several cell types in the innate immune system, such as macrophage.

Lipopolysaccharide (LPS) is the major component of the outer membrane of Gram-negative bacteria, which could activate the innate immune response in humans by stimulating inflammatory cytokine release such as tumornecrosis factor-alpha (TNFa) from monocytes and macrophages. Toll-like receptor 4 (TLR4) serves as a receptor for LPS in monocytes and macrophages. Dysregulation of TLR4 activation by LPS is responsible for chronic and acute inflammatory disorders that often cause dangerous disease like sepsis that still lacks specific pharmacological treatment. Previous studies confirmed that LPS induces Neu1 sialidase expression in macrophages, enhancing LPS/TLR4 receptor complex formation, NF-κB activation, and pro-inflammatory cytokine production. It is speculated that Neu1 sialidase could relocate to cell surface and cause desialylation of cell surface TLR4 and its activation. However, the molecular mechanisms related to LPS-activated sialidase expression and the desialylation in the LPS/TLR4 signaling pathway are still unclear. In the first chapter, we investigated sialylation status and sialidase espression, relocation and seceretion from monocytes and macrophages under LPS stimulation. We confirmed that the desialydation of LPS stimulated THP-1 monocytes and THP-1 macrophages are closely related to the enhanced sialidase Neu1 and Neu3 activity. In particular, we found the enhanced sialidase activities in the cell culture medium, which support sialidase was released from THP-1 monocytes and THP-1 macrophages upon LPS stimulation. The secreted sialidase could serve as exogenous sialidase to desialylate cell surface glycoconjugates. This is the first report about sialidase such as Neu1 and Neu3 expression, relocation, and secretion from monocytes and macrophages upon LPS stimulation, indicating that endogenous sialidase expression, relocation, and secretion are all involved in LPS/TLR4 signaling pathway and subsequent biological processes and cellular functions

In the second chapter, we did chemical modulation of desialylation of macrophages under LPS stimulation and its function. We characterized the sialylation status of LPS-stimulated THP-1 macrophages and their elastic properties and function as well. Specifically, we confirmed that sialylation status is closely related to macrophage elastic properties and is directly involved in macrophage functions. Further, we modulated macrophage sialylation status by feeding the cells with free sialic acid (Neu5Ac, Neu5Gc) and sialidase inhibitors and examined the impact on cellular elasticity and phagocytosis. The esults taken together suggest that although THP-1 macrophage stimulation with 100 ng/mL LPS did not affect their viability, it nevertheless induced a significant reduction in Sia levels (Neu5Ac & Neu5Gc) and elevation in α2,6-linkaged Sia expressions, which positively contributed to robust increases in cell elastic modulus, adhesion/tether forces and membrane tension, which in turn contributed to relatively higher phagocytosis activity.

Meanwhile, exogenous supplementation of Sias or Sia inhibitors to LPS-stimulated cells significantly and consistently reduced the measured biomechanical characteristics of these cells. Studies such as this unravel the mechanisms by which Sias modulate macrophage functions and establish the relationship between the biological, biomechanical, and functional outcomes of these cells when Sia levels are altered on their surfaces. Overall, the sialylation status is closely related to the cellular elasticity of macrophages.

Finally, we systematically examined the metabolic modification of cell surface sialoform of THP-1 macrophages with human and non-human Sia, Neu5Ac, Neu5Gc, and Neu5Gly. HPLC analysis identified the Sias incorporation in membrane glycoproteins and glycolipids. LC-MS/MS quantification confirmed the total incorporation of Sias. We found that human THP-1 macrophage cell lines intake Neu5Gc and Neu5Gly in a concentration-dependent way. These cell surface sialoform changes affect their specific lectin binding confirmed under flow cytometry and confocal imaging analysis. Also, the phagocytosis capability changed as compared to non-treated cells. Although the incorporation yield is low for THP-1 macrophages, metabolic incorporation of zwitterionic Sia for other cells needs to be investigated. It provides a new chemical functionality of glycan chain on the cells' cell surface and new functionality.

In summary, the sialylation and desialylation level of THP-1 monocyte and macrophage were quantified by the established LC-MS/MS method, and sialidases translocation was examined with Western Blotting. The secreted sialidases activity was enhanced under LPS stimulation, the localization and amount required to be detected in the future. Meanwhile, the THP-1 macrophage mechanical properties were detected through AFM, and it was found related to the sialylation levels of THP-1 macrophage. The zwitterionic Sis was used to metabolic remodel THP-1 macrophage; however, the incorporation is too tiny to alter the macrophage's function. Overall, a systematic recognition and modulation of sialylation and desialylation related to cellular property and function of macrophages will contribute to understanding specific signaling pathways in macrophages related to both physiological and pathological pathways and defining the phenotypes and elucidate functional diversity of macrophages. Further, it will provide novel mechanisms and approaches for diseases diagnosis and treatment.

CHAPTER VI

FUTURE DIRECTION

6.1 The Exocellular Level of Sialidase of THP-1 Monocytes Differentiation and LPS Stimulation

In this work, during the THP-1 monocyte differentiation and LPS stimulation, sialidase activities are significantly increased inside the cell culture medium. However, the type of the sialidase in the cell culture medium and the location of sialidases are still unknown. In the future, the ELISA will be used to detect sialidases in the cell culture medium. Meanwhile, the exo-vesicle will be isolated from the cell culture medium and identified by TEM and western blot of Biomarkers, and sialidases in the exo-vesicle could be identified by Western blot. The cell culture medium from the LPS stimulated THP-1 macrophage could be used to treat the other cells, such as fresh THP-1 monocyte or macrophage. And the total Sia will be tested in the fresh cells cell lysates and the cell culture medium.

6.2 The Expression and Localization of PPCA in THP-1 Monocytes and Macrophage during the Differentiation and LPS Stimulation

In this work, the Neu1 activity was significantly increased upon LPS stimulation, but the amount is not significantly increased. The reason that could cause the increased activity is the co-enzyme activation. The PPCA was reported as the co-enzyme of Neu1, which could be used to activate the function of sialidase. However, the PPCA level and location in the cells and the cell culture medium were not detected in THP-1 monocyte and macrophage upon LPS stimulation. In the future, the mRNA and protein level need to be detected by RT-PCR and Western blot. In addition, the location of PPCA and Neu1 need to be detected by co-localization of Microscopy.

6.3. Alternated the Macrophage Function by Introducing the Zwitterionic Sias

In this study, we systematically examined metabolic modification of cell surface sialoform of THP-1 macrophages with human and non-human Sia, Neu5Ac, Neu5Gc, and Neu5Gly, in the cell lysates and cell membrane, respectively. However, the expression level of Neu5Gly on THP-1 macrophages is significantly low. In the future, other zwitterionic Sia such as Neu5Amine and Neu5Ac9Amine will be introduced to the THP-1 macrophage. The presentation of zwitterionic Sias will be qualified in the cell lysates and on the cell membrane; meanwhile, the cell surface linkage change will be tested through lectin binding. The function and morphology should be examined; the phagocytosis could be detected by feeding THP-1 macrophage with FITC- labeled *E. Coil*, and morphology and properties could be tested by Confocal Microscopy and AFM.

6.4 Qualified the Desialylation of TLR4 upon LPS Stimulation.

The TLR4 is one of the well-known sialylated receptors found on the macrophage, the desialylation of TLR4 was reported related to the activation of the immune response. In the future, the level of sialylation of TLR4 will be qualified, and the function of the differential sialylation levels TLR4 need to be tested.

REFERENCES

[1] A. Varki, R. Schauer, Sialic Acids, in: nd, A. Varki, R.D. Cummings, J.D. Esko, H.H. Freeze, P. Stanley, C.R. Bertozzi, G.W. Hart, M.E. Etzler (Eds.) Essentials of Glycobiology, Place Published, 2009.

[2] A. Rambourg, C.P. Leblond, Electron microscope observations on the carbohydrate-rich cell coat present at the surface of cells in the rat, J Cell Biol, 32 (1967) 27-53.

[3] A. Varki, R.D. Cummings, J.D. Esko, P. Stanley, G.W. Hart, M. Aebi, A.G. Darvill, T. Kinoshita, N. Packer, J.H. Prestegard, R.L. Schnaar, P.H. Seeberger, Essentials of glycobiology, Third edition ed., Place Published, 2007.

[4] T. Angata, A. Varki, Chemical diversity in the sialic acids and related alpha-keto acids: an evolutionary perspective, Chem Rev, 102 (2002) 439-469.

[5] M. Cohen, A. Varki, The sialome--far more than the sum of its parts, OMICS, 14 (2010) 455-464.

[6] A. Harduin-Lepers, V. Vallejo-Ruiz, M.A. Krzewinski-Recchi, B. Samyn-Petit, S. Julien, P. Delannoy, The human sialyltransferase family, Biochimie, 83 (2001) 727-737.

[7] A. Varki, Glycan-based interactions involving vertebrate sialic-acid-recognizing proteins, Nature, 446 (2007) 1023-1029.

[8] P. Tangvoranuntakul, P. Gagneux, S. Diaz, M. Bardor, N. Varki, A. Varki, E. Muchmore, Human uptake and incorporation of an immunogenic nonhuman dietary sialic acid, Proc Natl Acad Sci U S A, 100 (2003) 12045-12050.

[9] A. Varki, Loss of N-glycolylneuraminic acid in humans: Mechanisms, consequences, and implications for hominid evolution, Am J Phys Anthropol, Suppl 33 (2001) 54-69.

[10] A. Varki, Sialic acids in human health and disease, Trends Mol Med, 14 (2008) 351-360.

[11] R. Schauer, Sialic acids as regulators of molecular and cellular interactions, Curr Opin Struct Biol, 19 (2009) 507-514.

[12] S.J. Moons, G.J. Adema, M.T. Derks, T.J. Boltje, C. Bull, Sialic acid glycoengineering using N-acetylmannosamine and sialic acid analogs, Glycobiology, 29 (2019) 433-445.

[13] J.C. Paulson, C. Rademacher, Glycan terminator, Nat Struct Mol Biol, 16 (2009) 1121-1122.

[14] R. Stasche, S. Hinderlich, C. Weise, K. Effertz, L. Lucka, P. Moormann, W. Reutter, A bifunctional enzyme catalyzes the first two steps in N-acetylneuraminic acid biosynthesis of rat liver. Molecular cloning and functional expression of UDP-N-acetyl-glucosamine 2-epimerase/N-acetylmannosamine kinase, J Biol Chem, 272 (1997) 24319-24324.

[15] R. Horstkorte, S. Nohring, N. Wiechens, M. Schwarzkopf, K. Danker, W. Reutter, L. Lucka, Tissue expression and amino acid sequence of murine UDP-N-acetylglucosamine-2-epimerase/N-acetylmannosamine kinase, Eur J Biochem, 260 (1999) 923-927.

[16] L. Lucka, M. Krause, K. Danker, W. Reutter, R. Horstkorte, Primary structure and expression analysis of human UDP-N-acetyl-glucosamine-2-epimerase/N-

acetylmannosamine kinase, the bifunctional enzyme in neuraminic acid biosynthesis, FEBS Lett, 454 (1999) 341-344.

[17] R. Seppala, V.P. Lehto, W.A. Gahl, Mutations in the human UDP-N-acetylglucosamine 2-epimerase gene define the disease sialuria and the allosteric site of the enzyme, Am J Hum Genet, 64 (1999) 1563-1569.

[18] N.M. Varki, A. Varki, Diversity in cell surface sialic acid presentations: implications for biology and disease, Lab Invest, 87 (2007) 851-857.

[19] G. Dekan, C. Gabel, M.G. Farquhar, Sulfate contributes to the negative charge of podocalyxin, the major sialoglycoprotein of the glomerular filtration slits, Proc Natl Acad Sci U S A, 88 (1991) 5398-5402.

[20] H. Gelberg, L. Healy, H. Whiteley, L.A. Miller, E. Vimr, In vivo enzymatic removal of alpha 2-->6-linked sialic acid from the glomerular filtration barrier results in podocyte charge alteration and glomerular injury, Lab Invest, 74 (1996) 907-920.

[21] B. Weinhold, R. Seidenfaden, I. Rockle, M. Muhlenhoff, F. Schertzinger, S. Conzelmann, J.D. Marth, R. Gerardy-Schahn, H. Hildebrandt, Genetic ablation of polysialic acid causes severe neurodevelopmental defects rescued by deletion of the neural cell adhesion molecule, J Biol Chem, 280 (2005) 42971-42977.

[22] C.P. Johnson, I. Fujimoto, U. Rutishauser, D.E. Leckband, Direct evidence that neural cell adhesion molecule (NCAM) polysialylation increases intermembrane repulsion and abrogates adhesion, J Biol Chem, 280 (2005) 137-145.

[23] A. El Maarouf, A.K. Petridis, U. Rutishauser, Use of polysialic acid in repair of the central nervous system, Proc Natl Acad Sci U S A, 103 (2006) 16989-16994.

[24] U. Rutishauser, Polysialic acid in the plasticity of the developing and adult vertebrate nervous system, Nat Rev Neurosci, 9 (2008) 26-35.

[25] G.V. Born, W. Palinski, Unusually high concentrations of sialic acids on the surface of vascular endothelia, Br J Exp Pathol, 66 (1985) 543-549.

[26] G. Ashwell, J. Harford, Carbohydrate-specific receptors of the liver, Annu Rev Biochem, 51 (1982) 531-554.

[27] P.H. Weigel, J.H. Yik, Glycans as endocytosis signals: the cases of the asialoglycoprotein and hyaluronan/chondroitin sulfate receptors, Biochim Biophys Acta, 1572 (2002) 341-363.

[28] J.C. Pickup, M.B. Mattock, M.A. Crook, G.D. Chusney, D. Burt, A.P. Fitzgerald, Serum sialic acid concentration and coronary heart disease in NIDDM, Diabetes Care, 18 (1995) 1100-1103.

[29] M.A. Crook, K. Earle, A. Morocutti, J. Yip, G. Viberti, J.C. Pickup, Serum sialic acid, a risk factor for cardiovascular disease, is increased in IDDM patients with microalbuminuria and clinical proteinuria, Diabetes Care, 17 (1994) 305-310.

[30] M. Ponnio, H. Alho, S.T. Nikkari, U. Olsson, U. Rydberg, P. Sillanaukee, Serum sialic acid in a random sample of the general population, Clin Chem, 45 (1999) 1842-1849.

[31] B. Afzali, R.S. Bakri, P. Bharma-Ariza, P.J. Lumb, N. Dalton, N.C. Turner, A.S. Wierzbicki, M.A. Crook, D.J. Goldsmith, Raised plasma total sialic acid levels are markers of cardiovascular disease in renal dialysis patients, J Nephrol, 16 (2003) 540-545.

[32] J.S. Millar, The sialylation of plasma lipoproteins, Atherosclerosis, 154 (2001) 1-13.

[33] V.V. Tertov, V.V. Kaplun, I.A. Sobenin, E.Y. Boytsova, N.V. Bovin, A.N. Orekhov, Human plasma trans-sialidase causes atherogenic modification of low density lipoprotein, Atherosclerosis, 159 (2001) 103-115.

[34] P. Gunnarsson, L. Levander, P. Pahlsson, M. Grenegard, The acute-phase protein alpha 1-acid glycoprotein (AGP) induces rises in cytosolic Ca2+ in neutrophil granulocytes via sialic acid binding immunoglobulin-like lectins (siglecs), FASEB J, 21 (2007) 4059-4069.

[35] F. Liang, V. Seyrantepe, K. Landry, R. Ahmad, A. Ahmad, N.M. Stamatos, A.V. Pshezhetsky, Monocyte differentiation up-regulates the expression of the lysosomal sialidase, Neu1, and triggers its targeting to the plasma membrane via major histocompatibility complex class II-positive compartments, J Biol Chem, 281 (2006) 27526-27538.

[36] X. Nan, I. Carubelli, N.M. Stamatos, Sialidase expression in activated human T lymphocytes influences production of IFN-gamma, J Leukoc Biol, 81 (2007) 284-296.

[37] S.A. Wuensch, R.Y. Huang, J. Ewing, X. Liang, J.T. Lau, Murine B cell differentiation is accompanied by programmed expression of multiple novel beta-galactoside alpha2, 6-sialyltransferase mRNA forms, Glycobiology, 10 (2000) 67-75.

[38] P.R. Crocker, J.C. Paulson, A. Varki, Siglecs and their roles in the immune system, Nat Rev Immunol, 7 (2007) 255-266.

[39] A. Varki, T. Angata, Siglecs--the major subfamily of I-type lectins, Glycobiology, 16 (2006) 1R-27R.

[40] C. Oetke, M.C. Vinson, C. Jones, P.R. Crocker, Sialoadhesin-deficient mice exhibit subtle changes in B- and T-cell populations and reduced immunoglobulin M levels, Mol Cell Biol, 26 (2006) 1549-1557.

[41] S. Tsuji, A.K. Datta, J.C. Paulson, Systematic nomenclature for sialyltransferases, Glycobiology, 6 (1996) v-vii.

[42] L. Wang, Y. Liu, L. Wu, X.L. Sun, Sialyltransferase inhibition and recent advances, Biochim Biophys Acta, 1864 (2016) 143-153.

[43] E. Monti, A. Preti, B. Venerando, G. Borsani, Recent development in mammalian sialidase molecular biology, Neurochem Res, 27 (2002) 649-663.

[44] T. Miyagi, K. Yamaguchi, Mammalian sialidases: physiological and pathological roles in cellular functions, Glycobiology, 22 (2012) 880-896.

[45] R.K.Y. Megumi Saito, Biochemistry and Function of Sialidases, Biochemistry and Function of Sialidases, (1995).

[46] D. Petit, E. Teppa, U. Cenci, S. Ball, A. Harduin-Lepers, Reconstruction of the sialylation pathway in the ancestor of eukaryotes, Sci Rep, 8 (2018) 2946.

[47] M.E. Tanner, The enzymes of sialic acid biosynthesis, Bioorg Chem, 33 (2005) 216-228.

[48] P. Maliekal, D. Vertommen, G. Delpierre, E. Van Schaftingen, Identification of the sequence encoding N-acetylneuraminate-9-phosphate phosphatase, Glycobiology, 16 (2006) 165-172.

[49] Y. Li, X. Chen, Sialic acid metabolism and sialyltransferases: natural functions and applications, Appl Microbiol Biotechnol, 94 (2012) 887-905.

[50] A.K. Munster-Kuhnel, J. Tiralongo, S. Krapp, B. Weinhold, V. Ritz-Sedlacek, U. Jacob, R. Gerardy-Schahn, Structure and function of vertebrate CMP-sialic acid synthetases, Glycobiology, 14 (2004) 43R-51R.

[51] L. Warren, H. Felsenfeld, The biosynthesis of sialic acids, J Biol Chem, 237 (1962) 1421-1431.

[52] S. Hinderlich, W. Weidemann, T. Yardeni, R. Horstkorte, M. Huizing, UDP-GlcNAc 2-Epimerase/ManNAc Kinase (GNE): A Master Regulator of Sialic Acid Synthesis, Top Curr Chem, 366 (2015) 97-137.

[53] L.R. Davies, A. Varki, Why Is N-Glycolylneuraminic Acid Rare in the Vertebrate Brain?, Top Curr Chem, 366 (2015) 31-54.

[54] C. Mandal, R. Schwartz-Albiez, R. Vlasak, Functions and Biosynthesis of O-Acetylated Sialic Acids, Top Curr Chem, 366 (2015) 1-30.

[55] P.A. De Bank, B. Kellam, D.A. Kendall, K.M. Shakesheff, Surface engineering of living myoblasts via selective periodate oxidation, Biotechnol Bioeng, 81 (2003) 800-808.

[56] Y. Zeng, T.N. Ramya, A. Dirksen, P.E. Dawson, J.C. Paulson, High-efficiency labeling of sialylated glycoproteins on living cells, Nat Methods, 6 (2009) 207-209.

[57] E. Monti, E. Bonten, A. D'Azzo, R. Bresciani, B. Venerando, G. Borsani, R. Schauer, G. Tettamanti, Sialidases in vertebrates: a family of enzymes tailored for several cell functions, Adv Carbohydr Chem Biochem, 64 (2010) 403-479.

[58] V.Y. Glanz, V.A. Myasoedova, A.V. Grechko, A.N. Orekhov, Sialidase activity in human pathologies, Eur J Pharmacol, 842 (2019) 345-350.

[59] E. Bonten, A. van der Spoel, M. Fornerod, G. Grosveld, A. d'Azzo, Characterization of human lysosomal neuraminidase defines the molecular basis of the metabolic storage disorder sialidosis, Genes Dev, 10 (1996) 3156-3169.

[60] P. Maurice, S. Baud, O.V. Bocharova, E.V. Bocharov, A.S. Kuznetsov, C. Kawecki, O. Bocquet, B. Romier, L. Gorisse, M. Ghirardi, L. Duca, S. Blaise, L. Martiny, M. Dauchez, R.G. Efremov, L. Debelle, New Insights into Molecular Organization of Human Neuraminidase-1: Transmembrane Topology and Dimerization Ability, Sci Rep, 6 (2016) 38363.

[61] L. Bao, L. Ding, J. Hui, H. Ju, A light-up imaging protocol for neutral pH-enhanced fluorescence detection of lysosomal neuraminidase activity in living cells, Chem Commun (Camb), 52 (2016) 12897-12900.

[62] L.M. Chavas, C. Tringali, P. Fusi, B. Venerando, G. Tettamanti, R. Kato, E. Monti, S. Wakatsuki, Crystal structure of the human cytosolic sialidase Neu2. Evidence for the dynamic nature of substrate recognition, J Biol Chem, 280 (2005) 469-475.

[63] E. Monti, A. Preti, E. Rossi, A. Ballabio, G. Borsani, Cloning and characterization of NEU2, a human gene homologous to rodent soluble sialidases, Genomics, 57 (1999) 137-143.

[64] T. Miyagi, T. Wada, A. Iwamatsu, K. Hata, Y. Yoshikawa, S. Tokuyama, M. Sawada, Molecular cloning and characterization of a plasma membrane-associated sialidase specific for gangliosides, J Biol Chem, 274 (1999) 5004-5011.

[65] E. Monti, M.T. Bassi, R. Bresciani, S. Civini, G.L. Croci, N. Papini, M. Riboni, G. Zanchetti, A. Ballabio, A. Preti, G. Tettamanti, B. Venerando, G. Borsani, Molecular cloning and characterization of NEU4, the fourth member of the human sialidase gene family, Genomics, 83 (2004) 445-453.

[66] M. Wei, P.G. Wang, Desialylation in physiological and pathological processes: New target for diagnostic and therapeutic development, Prog Mol Biol Transl Sci, 162 (2019) 25-57.

[67] R. Schauer, Sialic acids: fascinating sugars in higher animals and man, Zoology (Jena), 107 (2004) 49-64.

[68] K.E. Achyuthan, A.M. Achyuthan, Comparative enzymology, biochemistry and pathophysiology of human exo-alpha-sialidases (neuraminidases), Comp Biochem Physiol B Biochem Mol Biol, 129 (2001) 29-64.

[69] K.E. Lukong, V. Seyrantepe, K. Landry, S. Trudel, A. Ahmad, W.A. Gahl, S. Lefrancois, C.R. Morales, A.V. Pshezhetsky, Intracellular distribution of lysosomal sialidase is controlled by the internalization signal in its cytoplasmic tail, J Biol Chem, 276 (2001) 46172-46181.

[70] L. Duca, C. Blanchevoye, B. Cantarelli, C. Ghoneim, S. Dedieu, F. Delacoux, W. Hornebeck, A. Hinek, L. Martiny, L. Debelle, The elastin receptor complex transduces signals through the catalytic activity of its Neu-1 subunit, J Biol Chem, 282 (2007) 12484-12491.

[71] A. Rusciani, L. Duca, H. Sartelet, A. Chatron-Colliet, H. Bobichon, D. Ploton, R. Le Naour, S. Blaise, L. Martiny, L. Debelle, Elastin peptides signaling relies on

neuraminidase-1-dependent lactosylceramide generation, PLoS One, 5 (2010) e14010.

[72] A. Hinek, A.V. Pshezhetsky, M. von Itzstein, B. Starcher, Lysosomal sialidase (neuraminidase-1) is targeted to the cell surface in a multiprotein complex that facilitates elastic fiber assembly, J Biol Chem, 281 (2006) 3698-3710.

[73] S. Blaise, B. Romier, C. Kawecki, M. Ghirardi, F. Rabenoelina, S. Baud, L. Duca, P. Maurice, A. Heinz, C.E. Schmelzer, M. Tarpin, L. Martiny, C. Garbar, M. Dauchez, L. Debelle, V. Durlach, Elastin-derived peptides are new regulators of insulin resistance development in mice, Diabetes, 62 (2013) 3807-3816.

[74] L. Dridi, V. Seyrantepe, A. Fougerat, X. Pan, E. Bonneil, P. Thibault, A. Moreau, G.A. Mitchell, N. Heveker, C.W. Cairo, T. Issad, A. Hinek, A.V. Pshezhetsky, Positive regulation of insulin signaling by neuraminidase 1, Diabetes, 62 (2013) 2338-2346.

[75] T. Uemura, K. Shiozaki, K. Yamaguchi, S. Miyazaki, S. Satomi, K. Kato, H. Sakuraba, T. Miyagi, Contribution of sialidase NEU1 to suppression of metastasis of human colon cancer cells through desialylation of integrin beta4, Oncogene, 28 (2009) 1218-1229.

[76] S.R. Amith, P. Jayanth, S. Franchuk, T. Finlay, V. Seyrantepe, R. Beyaert, A.V. Pshezhetsky, M.R. Szewczuk, Neu1 desialylation of sialyl alpha-2,3-linked beta-galactosyl residues of TOLL-like receptor 4 is essential for receptor activation and cellular signaling, Cell Signal, 22 (2010) 314-324.

[77] P. Jayanth, S.R. Amith, K. Gee, M.R. Szewczuk, Neu1 sialidase and matrix metalloproteinase-9 cross-talk is essential for neurotrophin activation of Trk receptors and cellular signaling, Cell Signal, 22 (2010) 1193-1205.

[78] A. Hinek, T.D. Bodnaruk, S. Bunda, Y. Wang, K. Liu, Neuraminidase-1, a subunit of the cell surface elastin receptor, desialylates and functionally inactivates adjacent receptors interacting with the mitogenic growth factors PDGF-BB and IGF-2, Am J Pathol, 173 (2008) 1042-1056.

[79] E.P. Lillehoj, S.W. Hyun, C. Feng, L. Zhang, A. Liu, W. Guang, C. Nguyen, I.G. Luzina, S.P. Atamas, A. Passaniti, W.S. Twaddell, A.C. Puche, L.X. Wang, A.S. Cross, S.E. Goldblum, NEU1 sialidase expressed in human airway epithelia regulates epidermal growth factor receptor (EGFR) and MUC1 protein signaling, J Biol Chem, 287 (2012) 8214-8231.

[80] C. Lee, A. Liu, A. Miranda-Ribera, S.W. Hyun, E.P. Lillehoj, A.S. Cross, A. Passaniti, P.R. Grimm, B.Y. Kim, P.A. Welling, J.A. Madri, H.M. DeLisser, S.E. Goldblum, NEU1 sialidase regulates the sialylation state of CD31 and disrupts CD31-driven capillary-like tube formation in human lung microvascular endothelia, J Biol Chem, 289 (2014) 9121-9135.

[81] E.J. Bonten, I. Annunziata, A. d'Azzo, Lysosomal multienzyme complex: pros and cons of working together, Cell Mol Life Sci, 71 (2014) 2017-2032.

[82] E.J. Bonten, Y. Campos, V. Zaitsev, A. Nourse, B. Waddell, W. Lewis, G. Taylor, A. d'Azzo, Heterodimerization of the sialidase NEU1 with the chaperone protective protein/cathepsin A prevents its premature oligomerization, J Biol Chem, 284 (2009) 28430-28441.

[83] A. van der Spoel, E. Bonten, A. d'Azzo, Transport of human lysosomal neuraminidase to mature lysosomes requires protective protein/cathepsin A, EMBO J, 17 (1998) 1588-1597.

[84] G. Yogalingam, E.J. Bonten, D. van de Vlekkert, H. Hu, S. Moshiach, S.A. Connell, A. d'Azzo, Neuraminidase 1 is a negative regulator of lysosomal exocytosis, Dev Cell, 15 (2008) 74-86.

[85] C. Huynh, D. Roth, D.M. Ward, J. Kaplan, N.W. Andrews, Defective lysosomal exocytosis and plasma membrane repair in Chediak-Higashi/beige cells, Proc Natl Acad Sci U S A, 101 (2004) 16795-16800.

[86] J.K. Jaiswal, N.W. Andrews, S.M. Simon, Membrane proximal lysosomes are the major vesicles responsible for calcium-dependent exocytosis in nonsecretory cells, J Cell Biol, 159 (2002) 625-635.

[87] P.L. McNeil, R.A. Steinhardt, Plasma membrane disruption: repair, prevention, adaptation, Annu Rev Cell Dev Biol, 19 (2003) 697-731.

[88] A. Reddy, E.V. Caler, N.W. Andrews, Plasma membrane repair is mediated by Ca(2+)-regulated exocytosis of lysosomes, Cell, 106 (2001) 157-169.

[89] D. Roy, D.R. Liston, V.J. Idone, A. Di, D.J. Nelson, C. Pujol, J.B. Bliska, S. Chakrabarti, N.W. Andrews, A process for controlling intracellular bacterial infections induced by membrane injury, Science, 304 (2004) 1515-1518.

[90] N. de Geest, E. Bonten, L. Mann, J. de Sousa-Hitzler, C. Hahn, A. d'Azzo, Systemic and neurologic abnormalities distinguish the lysosomal disorders sialidosis and galactosialidosis in mice, Hum Mol Genet, 11 (2002) 1455-1464.

[91] P.E. Kima, B. Burleigh, N.W. Andrews, Surface-targeted lysosomal membrane glycoprotein-1 (Lamp-1) enhances lysosome exocytosis and cell invasion by Trypanosoma cruzi, Cell Microbiol, 2 (2000) 477-486.

[92] E. Machado, S. White-Gilbertson, D. van de Vlekkert, L. Janke, S. Moshiach, Y. Campos, D. Finkelstein, E. Gomero, R. Mosca, X. Qiu, C.L. Morton, I. Annunziata, A. d'Azzo, Regulated lysosomal exocytosis mediates cancer progression, Sci Adv, 1 (2015) e1500603.

[93] R. Schauer, Achievements and challenges of sialic acid research, Glycoconj J, 17 (2000) 485-499.

[94] E.U. Bagriacik, K.S. Miller, Cell surface sialic acid and the regulation of immune cell interactions: the neuraminidase effect reconsidered, Glycobiology, 9 (1999) 267-275.

[95] M. Frohman, C. Cowing, Presentation of antigen by B cells: functional dependence on radiation dose, interleukins, cellular activation, and differential glycosylation, J Immunol, 134 (1985) 2269-2275.

[96] E.E. Eynon, D.C. Parker, Small B cells as antigen-presenting cells in the induction of tolerance to soluble protein antigens, J Exp Med, 175 (1992) 131-138.

[97] E.J. Fuchs, P. Matzinger, B cells turn off virgin but not memory T cells, Science, 258 (1992) 1156-1159.

[98] X.P. Chen, E.Y. Enioutina, R.A. Daynes, The control of IL-4 gene expression in activated murine T lymphocytes: a novel role for neu-1 sialidase, J Immunol, 158 (1997) 3070-3080.

[99] N.F. Landolfi, R.G. Cook, Activated T-lymphocytes express class I molecules which are hyposialylated compared to other lymphocyte populations, Mol Immunol, 23 (1986) 297-309.

[100] N. Yamamoto, R. Kumashiro, Conversion of vitamin D3 binding protein (group-specific component) to a macrophage activating factor by the stepwise action of beta-galactosidase of B cells and sialidase of T cells, J Immunol, 151 (1993) 2794-2802.

[101] D. Wang, E. Ozhegov, L. Wang, A. Zhou, H. Nie, Y. Li, X.L. Sun, Sialylation and desialylation dynamics of monocytes upon differentiation and polarization to macrophages, Glycoconj J, 33 (2016) 725-733.

[102] N.M. Stamatos, F. Liang, X. Nan, K. Landry, A.S. Cross, L.X. Wang, A.V. Pshezhetsky, Differential expression of endogenous sialidases of human monocytes during cellular differentiation into macrophages, FEBS J, 272 (2005) 2545-2556.

[103] V. Seyrantepe, A. Iannello, F. Liang, E. Kanshin, P. Jayanth, S. Samarani, M.R. Szewczuk, A. Ahmad, A.V. Pshezhetsky, Regulation of phagocytosis in macrophages by neuraminidase 1, J Biol Chem, 285 (2010) 206-215.

[104] S.R. Amith, P. Jayanth, S. Franchuk, S. Siddiqui, V. Seyrantepe, K. Gee, S. Basta, R. Beyaert, A.V. Pshezhetsky, M.R. Szewczuk, Dependence of pathogen molecule-induced toll-like receptor activation and cell function on Neu1 sialidase, Glycoconj J, 26 (2009) 1197-1212.

[105] I. Sieve, M. Ricke-Hoch, M. Kasten, K. Battmer, B. Stapel, C.S. Falk, M.S. Leisegang, A. Haverich, M. Scherr, D. Hilfiker-Kleiner, A positive feedback loop

between IL-1beta, LPS and NEU1 may promote atherosclerosis by enhancing a pro-inflammatory state in monocytes and macrophages, Vascul Pharmacol, 103-105 (2018) 16-28.

[106] G.Y. Chen, N.K. Brown, W. Wu, Z. Khedri, H. Yu, X. Chen, D. van de Vlekkert, A. D'Azzo, P. Zheng, Y. Liu, Broad and direct interaction between TLR and Siglec families of pattern recognition receptors and its regulation by Neu1, Elife, 3 (2014) e04066.

[107] B. Starcher, A. d'Azzo, P.W. Keller, G.K. Rao, D. Nadarajah, A. Hinek, Neuraminidase-1 is required for the normal assembly of elastic fibers, Am J Physiol Lung Cell Mol Physiol, 295 (2008) L637-647.

[108] H. Morreau, N.J. Galjart, R. Willemsen, N. Gillemans, X.Y. Zhou, A. d'Azzo, Human lysosomal protective protein. Glycosylation, intracellular transport, and association with beta-galactosidase in the endoplasmic reticulum, J Biol Chem, 267 (1992) 17949-17956.

[109] B. Guertl, C. Noehammer, G. Hoefler, Metabolic cardiomyopathies, Int J Exp Pathol, 81 (2000) 349-372.

[110] T. Miyagi, K. Sato, K. Hata, S. Taniguchi, Metastatic potential of transformed rat 3Y1 cell lines is inversely correlated with lysosomal-type sialidase activity, FEBS Lett, 349 (1994) 255-259.

[111] M. Sawada, S. Moriya, S. Saito, R. Shineha, S. Satomi, T. Yamori, T. Tsuruo, R. Kannagi, T. Miyagi, Reduced sialidase expression in highly metastatic variants of mouse colon adenocarcinoma 26 and retardation of their metastatic ability by sialidase overexpression, Int J Cancer, 97 (2002) 180-185.

[112] T. Kato, Y. Wang, K. Yamaguchi, C.M. Milner, R. Shineha, S. Satomi, T. Miyagi, Overexpression of lysosomal-type sialidase leads to suppression of metastasis associated with reversion of malignant phenotype in murine B16 melanoma cells, Int J Cancer, 92 (2001) 797-804.

[113] L.R. Ren, L.P. Zhang, S.Y. Huang, Y.F. Zhu, W.J. Li, S.Y. Fang, L. Shen, Y.L. Gao, Effects of sialidase NEU1 siRNA on proliferation, apoptosis, and invasion in human ovarian cancer, Mol Cell Biochem, 411 (2016) 213-219.

[114] F. Haxho, R.J. Neufeld, M.R. Szewczuk, Neuraminidase-1: a novel therapeutic target in multistage tumorigenesis, Oncotarget, 7 (2016) 40860-40881.

[115] A.M. Gilmour, S. Abdulkhalek, T.S. Cheng, F. Alghamdi, P. Jayanth, L.K. O'Shea, O. Geen, L.A. Arvizu, M.R. Szewczuk, A novel epidermal growth factor receptor-signaling platform and its targeted translation in pancreatic cancer, Cell Signal, 25 (2013) 2587-2603.

[116] L.K. O'Shea, S. Abdulkhalek, S. Allison, R.J. Neufeld, M.R. Szewczuk, Therapeutic targeting of Neu1 sialidase with oseltamivir phosphate (Tamiflu(R)) disables cancer cell survival in human pancreatic cancer with acquired chemoresistance, Onco Targets Ther, 7 (2014) 117-134.

[117] T. Miyagi, K. Konno, Y. Emori, H. Kawasaki, K. Suzuki, A. Yasui, S. Tsuik, Molecular cloning and expression of cDNA encoding rat skeletal muscle cytosolic sialidase, J Biol Chem, 268 (1993) 26435-26440.

[118] J. Ferrari, R. Harris, T.G. Warner, Cloning and expression of a soluble sialidase from Chinese hamster ovary cells: sequence alignment similarities to bacterial sialidases, Glycobiology, 4 (1994) 367-373.

[119] C.L. Fronda, G. Zeng, L. Gao, R.K. Yu, Molecular cloning and expression of mouse brain sialidase, Biochem Biophys Res Commun, 258 (1999) 727-731.

[120] T. Hasegawa, K. Yamaguchi, T. Wada, A. Takeda, Y. Itoyama, T. Miyagi, Molecular cloning of mouse ganglioside sialidase and its increased expression in Neuro2a cell differentiation, J Biol Chem, 275 (2000) 8007-8015.

[121] K. Kotani, A. Kuroiwa, T. Saito, Y. Matsuda, T. Koda, S. Kijimoto-Ochiai, Cloning, chromosomal mapping, and characteristic 5'-UTR sequence of murine cytosolic sialidase, Biochem Biophys Res Commun, 286 (2001) 250-258.

[122] C. Tringali, N. Papini, P. Fusi, G. Croci, G. Borsani, A. Preti, P. Tortora, G. Tettamanti, B. Venerando, E. Monti, Properties of recombinant human cytosolic sialidase HsNEU2. The enzyme hydrolyzes monomerically dispersed GM1 ganglioside molecules, J Biol Chem, 279 (2004) 3169-3179.

[123] K. Sato, T. Miyagi, Involvement of an endogenous sialidase in skeletal muscle cell differentiation, Biochem Biophys Res Commun, 221 (1996) 826-830.

[124] A. Fanzani, R. Giuliani, F. Colombo, D. Zizioli, M. Presta, A. Preti, S. Marchesini, Overexpression of cytosolic sialidase Neu2 induces myoblast differentiation in C2C12 cells, FEBS Lett, 547 (2003) 183-188.

[125] N. Suzuki, M. Aoki, Y. Hinuma, T. Takahashi, Y. Onodera, A. Ishigaki, M. Kato, H. Warita, M. Tateyama, Y. Itoyama, Expression profiling with progression of dystrophic change in dysferlin-deficient mice (SJL), Neurosci Res, 52 (2005) 47-60.

[126] A. Fanzani, F. Colombo, R. Giuliani, A. Preti, S. Marchesini, Cytosolic sialidase Neu2 upregulation during PC12 cells differentiation, FEBS Lett, 566 (2004) 178-182.

[127] A. Fanzani, F. Colombo, R. Giuliani, A. Preti, S. Marchesini, Insulin-like growth factor 1 signaling regulates cytosolic sialidase Neu2 expression during myoblast differentiation and hypertrophy, FEBS J, 273 (2006) 3709-3721.

[128] A. Fanzani, R. Giuliani, F. Colombo, S. Rossi, E. Stoppani, W. Martinet, A. Preti, S. Marchesini, The enzymatic activity of sialidase Neu2 is inversely regulated during in vitro myoblast hypertrophy and atrophy, Biochem Biophys Res Commun, 370 (2008) 376-381.

[129] T. Miyagi, K. Takahashi, K. Hata, K. Shiozaki, K. Yamaguchi, Sialidase significance for cancer progression, Glycoconj J, 29 (2012) 567-577.

[130] E.J. Meuillet, R. Kroes, H. Yamamoto, T.G. Warner, J. Ferrari, B. Mania-Farnell, D. George, A. Rebbaa, J.R. Moskal, E.G. Bremer, Sialidase gene transfection enhances epidermal growth factor receptor activity in an epidermoid carcinoma cell line, A431, Cancer Res, 59 (1999) 234-240.

[131] C. Tringali, B. Lupo, L. Anastasia, N. Papini, E. Monti, R. Bresciani, G. Tettamanti, B. Venerando, Expression of sialidase Neu2 in leukemic K562 cells induces apoptosis by impairing Bcr-Abl/Src kinases signaling, J Biol Chem, 282 (2007) 14364-14372.

[132] K. Hata, K. Koseki, K. Yamaguchi, S. Moriya, Y. Suzuki, S. Yingsakmongkon, G. Hirai, M. Sodeoka, M. von Itzstein, T. Miyagi, Limited inhibitory effects of

oseltamivir and zanamivir on human sialidases, Antimicrob Agents Chemother, 52 (2008) 3484-3491.

[133] T. Wada, Y. Yoshikawa, S. Tokuyama, M. Kuwabara, H. Akita, T. Miyagi, Cloning, expression, and chromosomal mapping of a human ganglioside sialidase, Biochem Biophys Res Commun, 261 (1999) 21-27.

[134] E. Monti, M.T. Bassi, N. Papini, M. Riboni, M. Manzoni, B. Venerando, G. Croci, A. Preti, A. Ballabio, G. Tettamanti, G. Borsani, Identification and expression of NEU3, a novel human sialidase associated to the plasma membrane, Biochem J, 349 (2000) 343-351.

[135] T. Hasegawa, C. Feijoo Carnero, T. Wada, Y. Itoyama, T. Miyagi, Differential expression of three sialidase genes in rat development, Biochem Biophys Res Commun, 280 (2001) 726-732.

[136] N. Papini, L. Anastasia, C. Tringali, G. Croci, R. Bresciani, K. Yamaguchi, T. Miyagi, A. Preti, A. Prinetti, S. Prioni, S. Sonnino, G. Tettamanti, B. Venerando, E. Monti, The plasma membrane-associated sialidase MmNEU3 modifies the ganglioside pattern of adjacent cells supporting its involvement in cell-to-cell interactions, J Biol Chem, 279 (2004) 16989-16995.

[137] K. Yamaguchi, K. Hata, T. Wada, S. Moriya, T. Miyagi, Epidermal growth factor-induced mobilization of a ganglioside-specific sialidase (NEU3) to membrane ruffles, Biochem Biophys Res Commun, 346 (2006) 484-490.

[138] S. Proshin, K. Yamaguchi, T. Wada, T. Miyagi, Modulation of neuritogenesis by ganglioside-specific sialidase (Neu 3) in human neuroblastoma NB-1 cells, Neurochem Res, 27 (2002) 841-846.

[139] J.A. Rodriguez, E. Piddini, T. Hasegawa, T. Miyagi, C.G. Dotti, Plasma membrane ganglioside sialidase regulates axonal growth and regeneration in hippocampal neurons in culture, J Neurosci, 21 (2001) 8387-8395.

[140] J.S. Da Silva, T. Hasegawa, T. Miyagi, C.G. Dotti, J. Abad-Rodriguez, Asymmetric membrane ganglioside sialidase activity specifies axonal fate, Nat Neurosci, 8 (2005) 606-615.

[141] D. Kalka, C. von Reitzenstein, J. Kopitz, M. Cantz, The plasma membrane ganglioside sialidase cofractionates with markers of lipid rafts, Biochem Biophys Res Commun, 283 (2001) 989-993.

[142] Y. Wang, K. Yamaguchi, T. Wada, K. Hata, X. Zhao, T. Fujimoto, T. Miyagi, A close association of the ganglioside-specific sialidase Neu3 with caveolin in membrane microdomains, J Biol Chem, 277 (2002) 26252-26259.

[143] L. Veracini, V. Simon, V. Richard, B. Schraven, V. Horejsi, S. Roche, C. Benistant, The Csk-binding protein PAG regulates PDGF-induced Src mitogenic signaling via GM1, J Cell Biol, 182 (2008) 603-614.

[144] E. Odintsova, T.D. Butters, E. Monti, H. Sprong, G. van Meer, F. Berditchevski, Gangliosides play an important role in the organization of CD82-enriched microdomains, Biochem J, 400 (2006) 315-325.

[145] X. Wang, Z. Rahman, P. Sun, E. Meuillet, D. George, E.G. Bremer, A. Al-Qamari, A.S. Paller, Ganglioside modulates ligand binding to the epidermal growth factor receptor, J Invest Dermatol, 116 (2001) 69-76.

[146] Y. Kawakami, K. Kawakami, W.F. Steelant, M. Ono, R.C. Baek, K. Handa, D.A. Withers, S. Hakomori, Tetraspanin CD9 is a "proteolipid," and its interaction with

alpha 3 integrin in microdomain is promoted by GM3 ganglioside, leading to inhibition of laminin-5-dependent cell motility, J Biol Chem, 277 (2002) 34349-34358.

[147] E.A. Miljan, E.G. Bremer, Regulation of growth factor receptors by gangliosides, Sci STKE, 2002 (2002) re15.

[148] J. Kopitz, C. von Reitzenstein, C. Muhl, M. Cantz, Role of plasma membrane ganglioside sialidase of human neuroblastoma cells in growth control and differentiation, Biochem Biophys Res Commun, 199 (1994) 1188-1193.

[149] S. Kappagantula, M.R. Andrews, M. Cheah, J. Abad-Rodriguez, C.G. Dotti, J.W. Fawcett, Neu3 sialidase-mediated ganglioside conversion is necessary for axon regeneration and is blocked in CNS axons, J Neurosci, 34 (2014) 2477-2492.

[150] A.A. Vyas, H.V. Patel, S.E. Fromholt, M. Heffer-Lauc, K.A. Vyas, J. Dang, M. Schachner, R.L. Schnaar, Gangliosides are functional nerve cell ligands for myelin-associated glycoprotein (MAG), an inhibitor of nerve regeneration, Proc Natl Acad Sci U S A, 99 (2002) 8412-8417.

[151] A. Mountney, M.R. Zahner, I. Lorenzini, M. Oudega, L.P. Schramm, R.L. Schnaar, Sialidase enhances recovery from spinal cord contusion injury, Proc Natl Acad Sci U S A, 107 (2010) 11561-11566.

[152] L. Anastasia, N. Papini, F. Colazzo, G. Palazzolo, C. Tringali, L. Dileo, M. Piccoli, E. Conforti, C. Sitzia, E. Monti, M. Sampaolesi, G. Tettamanti, B. Venerando, NEU3 sialidase strictly modulates GM3 levels in skeletal myoblasts C2C12 thus favoring their differentiation and protecting them from apoptosis, J Biol Chem, 283 (2008) 36265-36271.

[153] N. Papini, L. Anastasia, C. Tringali, L. Dileo, I. Carubelli, M. Sampaolesi, E. Monti, G. Tettamanti, B. Venerando, MmNEU3 sialidase over-expression in C2C12 myoblasts delays differentiation and induces hypertrophic myotube formation, J Cell Biochem, 113 (2012) 2967-2978.

[154] R. Scaringi, M. Piccoli, N. Papini, F. Cirillo, E. Conforti, S. Bergante, C. Tringali, A. Garatti, C. Gelfi, B. Venerando, L. Menicanti, G. Tettamanti, L. Anastasia, NEU3 sialidase is activated under hypoxia and protects skeletal muscle cells from apoptosis through the activation of the epidermal growth factor receptor signaling pathway and the hypoxia-inducible factor (HIF)-1alpha, J Biol Chem, 288 (2013) 3153-3162.

[155] Y. Kakugawa, T. Wada, K. Yamaguchi, H. Yamanami, K. Ouchi, I. Sato, T. Miyagi, Up-regulation of plasma membrane-associated ganglioside sialidase (Neu3) in human colon cancer and its involvement in apoptosis suppression, Proc Natl Acad Sci U S A, 99 (2002) 10718-10723.

[156] T. Wada, K. Hata, K. Yamaguchi, K. Shiozaki, K. Koseki, S. Moriya, T. Miyagi, A crucial role of plasma membrane-associated sialidase in the survival of human cancer cells, Oncogene, 26 (2007) 2483-2490.

[157] T. Miyagi, T. Wada, K. Yamaguchi, K. Hata, K. Shiozaki, Plasma membrane-associated sialidase as a crucial regulator of transmembrane signalling, J Biochem, 144 (2008) 279-285.

[158] K. Yamaguchi, K. Hata, K. Koseki, K. Shiozaki, H. Akita, T. Wada, S. Moriya, T. Miyagi, Evidence for mitochondrial localization of a novel human sialidase (NEU4), Biochem J, 390 (2005) 85-93.

[159] A. Bigi, L. Morosi, C. Pozzi, M. Forcella, G. Tettamanti, B. Venerando, E. Monti, P. Fusi, Human sialidase NEU4 long and short are extrinsic proteins bound to outer mitochondrial membrane and the endoplasmic reticulum, respectively, Glycobiology, 20 (2010) 148-157.

[160] K. Shiozaki, K. Koseki, K. Yamaguchi, M. Shiozaki, H. Narimatsu, T. Miyagi, Developmental change of sialidase neu4 expression in murine brain and its involvement in the regulation of neuronal cell differentiation, J Biol Chem, 284 (2009) 21157-21164.

[161] V. Seyrantepe, M. Canuel, S. Carpentier, K. Landry, S. Durand, F. Liang, J. Zeng, A. Caqueret, R.A. Gravel, S. Marchesini, C. Zwingmann, J. Michaud, C.R. Morales, T. Levade, A.V. Pshezhetsky, Mice deficient in Neu4 sialidase exhibit abnormal ganglioside catabolism and lysosomal storage, Hum Mol Genet, 17 (2008) 1556-1568.

[162] K. Takahashi, J. Mitoma, M. Hosono, K. Shiozaki, C. Sato, K. Yamaguchi, K. Kitajima, H. Higashi, K. Nitta, H. Shima, T. Miyagi, Sialidase NEU4 hydrolyzes polysialic acids of neural cell adhesion molecules and negatively regulates neurite formation by hippocampal neurons, J Biol Chem, 287 (2012) 14816-14826.

[163] T. Hasegawa, N. Sugeno, A. Takeda, M. Matsuzaki-Kobayashi, A. Kikuchi, K. Furukawa, T. Miyagi, Y. Itoyama, Role of Neu4L sialidase and its substrate ganglioside GD3 in neuronal apoptosis induced by catechol metabolites, FEBS Lett, 581 (2007) 406-412.

[164] F. Malisan, R. Testi, GD3 ganglioside and apoptosis, Biochim Biophys Acta, 1585 (2002) 179-187.

[165] R. Kleene, M. Schachner, Glycans and neural cell interactions, Nat Rev Neurosci, 5 (2004) 195-208.

[166] E.M. Comelli, M. Amado, S.R. Lustig, J.C. Paulson, Identification and expression of Neu4, a novel murine sialidase, Gene, 321 (2003) 155-161.

[167] H. Yamanami, K. Shiozaki, T. Wada, K. Yamaguchi, T. Uemura, Y. Kakugawa, T. Hujiya, T. Miyagi, Down-regulation of sialidase NEU4 may contribute to invasive properties of human colon cancers, Cancer Sci, 98 (2007) 299-307.

[168] K. Shiozaki, K. Yamaguchi, K. Takahashi, S. Moriya, T. Miyagi, Regulation of sialyl Lewis antigen expression in colon cancer cells by sialidase NEU4, J Biol Chem, 286 (2011) 21052-21061.

[169] V. Seyrantepe, K. Landry, S. Trudel, J.A. Hassan, C.R. Morales, A.V. Pshezhetsky, Neu4, a novel human lysosomal lumen sialidase, confers normal phenotype to sialidosis and galactosialidosis cells, J Biol Chem, 279 (2004) 37021-37029.

[170] V. Seyrantepe, P. Lema, A. Caqueret, L. Dridi, S. Bel Hadj, S. Carpentier, F. Boucher, T. Levade, L. Carmant, R.A. Gravel, E. Hamel, P. Vachon, G. Di Cristo, J.L. Michaud, C.R. Morales, A.V. Pshezhetsky, Mice doubly-deficient in lysosomal hexosaminidase A and neuraminidase 4 show epileptic crises and rapid neuronal loss, PLoS Genet, 6 (2010) e1001118.

[171] M.M. Fuster, J.D. Esko, The sweet and sour of cancer: glycans as novel therapeutic targets, Nat Rev Cancer, 5 (2005) 526-542.

[172] O.M. Pearce, H. Laubli, Sialic acids in cancer biology and immunity, Glycobiology, 26 (2016) 111-128.

[173] G. Lindberg, G.A. Eklund, B. Gullberg, L. Rastam, Serum sialic acid concentration and cardiovascular mortality, BMJ, 302 (1991) 143-146.

[174] A. Vaya, C. Falco, E. Reganon, V. Vila, V. Martinez-Sales, D. Corella, M.T. Contreras, J. Aznar, Influence of plasma and erythrocyte factors on red blood cell aggregation in survivors of acute myocardial infarction, Thromb Haemost, 91 (2004) 354-359.

[175] V.A. Hanson, Jr., S.A. Landaw, M. Flashner, S.D. Wax, W.R. Webb, Sialic acid-depleted red cells following acute myocardial infarction, Am Heart J, 99 (1980) 483-486.

[176] P. Resnitzky, A. Yaari, D. Danon, Biophysical characteristics of erythrocytes during acute myocardial infarction and venous thrombosis, Thromb Diath Haemorrh, 28 (1972) 524-532.

[177] C. Bull, M.H. den Brok, G.J. Adema, Sweet escape: sialic acids in tumor immune evasion, Biochim Biophys Acta, 1846 (2014) 238-246.

[178] K.J. Malmberg, M. Carlsten, A. Bjorklund, E. Sohlberg, Y.T. Bryceson, H.G. Ljunggren, Natural killer cell-mediated immunosurveillance of human cancer, Semin Immunol, 31 (2017) 20-29.

[179] A.F. Swindall, A.I. Londono-Joshi, M.J. Schultz, N. Fineberg, D.J. Buchsbaum, S.L. Bellis, ST6Gal-I protein expression is upregulated in human epithelial tumors and correlates with stem cell markers in normal tissues and colon cancer cell lines, Cancer Res, 73 (2013) 2368-2378.

[180] M. Perdicchio, J.M. Ilarregui, M.I. Verstege, L.A. Cornelissen, S.T. Schetters, S. Engels, M. Ambrosini, H. Kalay, H. Veninga, J.M. den Haan, L.A. van Berkel,

J.N. Samsom, P.R. Crocker, T. Sparwasser, L. Berod, J.J. Garcia-Vallejo, Y. van Kooyk, W.W. Unger, Sialic acid-modified antigens impose tolerance via inhibition of T-cell proliferation and de novo induction of regulatory T cells, Proc Natl Acad Sci U S A, 113 (2016) 3329-3334.

[181] J. Lu, J. Gu, Significance of beta-Galactoside alpha2,6 Sialyltranferase 1 in Cancers, Molecules, 20 (2015) 7509-7527.

[182] M. Amano, M. Galvan, J. He, L.G. Baum, The ST6Gal I sialyltransferase selectively modifies N-glycans on CD45 to negatively regulate galectin-1-induced CD45 clustering, phosphatase modulation, and T cell death, J Biol Chem, 278 (2003) 7469-7475.

[183] Y. Zhuo, R. Chammas, S.L. Bellis, Sialylation of beta1 integrins blocks cell adhesion to galectin-3 and protects cells against galectin-3-induced apoptosis, J Biol Chem, 283 (2008) 22177-22185.

[184] J. Hirabayashi, T. Hashidate, Y. Arata, N. Nishi, T. Nakamura, M. Hirashima, T. Urashima, T. Oka, M. Futai, W.E. Muller, F. Yagi, K. Kasai, Oligosaccharide specificity of galectins: a search by frontal affinity chromatography, Biochim Biophys Acta, 1572 (2002) 232-254.

[185] A.G. Blidner, S.P. Mendez-Huergo, A.J. Cagnoni, G.A. Rabinovich, Re-wiring regulatory cell networks in immunity by galectin-glycan interactions, FEBS Lett, 589 (2015) 3407-3418.

[186] J. Burchell, R. Poulsom, A. Hanby, C. Whitehouse, L. Cooper, H. Clausen, D. Miles, J. Taylor-Papadimitriou, An alpha2,3 sialyltransferase (ST3Gal I) is elevated in primary breast carcinomas, Glycobiology, 9 (1999) 1307-1311.

[187] M. Perez-Garay, B. Arteta, L. Pages, R. de Llorens, C. de Bolos, F. Vidal-Vanaclocha, R. Peracaula, alpha2,3-sialyltransferase ST3Gal III modulates pancreatic cancer cell motility and adhesion in vitro and enhances its metastatic potential in vivo, PLoS One, 5 (2010).

[188] D.S. Raclawska, F. Ttofali, A.A. Fletcher, D.N. Harper, B.S. Bochner, W.J. Janssen, C.M. Evans, Mucins and Their Sugars. Critical Mediators of Hyperreactivity and Inflammation, Ann Am Thorac Soc, 13 Suppl 1 (2016) S98-99.

[189] S. Gretschel, W. Haensch, P.M. Schlag, W. Kemmner, Clinical relevance of sialyltransferases ST6GAL-I and ST3GAL-III in gastric cancer, Oncology, 65 (2003) 139-145.

[190] S.V. Glavey, S. Manier, A. Natoni, A. Sacco, M. Moschetta, M.R. Reagan, L.S. Murillo, I. Sahin, P. Wu, Y. Mishima, Y. Zhang, W. Zhang, Y. Zhang, G. Morgan, L. Joshi, A.M. Roccaro, I.M. Ghobrial, M.E. O'Dwyer, The sialyltransferase ST3GAL6 influences homing and survival in multiple myeloma, Blood, 124 (2014) 1765-1776.

[191] M. Saito, H. Okayama, T. Yoshii, H. Higashi, H. Morioka, G. Hiasa, T. Sumimoto, S. Inaba, K. Nishimura, K. Inoue, A. Ogimoto, Y. Shigematsu, M. Hamada, J. Higaki, Clinical significance of global two-dimensional strain as a surrogate parameter of myocardial fibrosis and cardiac events in patients with hypertrophic cardiomyopathy, Eur Heart J Cardiovasc Imaging, 13 (2012) 617-623.

[192] W.H. Yang, C. Nussbaum, P.K. Grewal, J.D. Marth, M. Sperandio, Coordinated roles of ST3Gal-VI and ST3Gal-IV sialyltransferases in the synthesis of selectin ligands, Blood, 120 (2012) 1015-1026.

[193] A.S. Carvalho, A. Harduin-Lepers, A. Magalhaes, E. Machado, N. Mendes, L.T. Costa, R. Matthiesen, R. Almeida, J. Costa, C.A. Reis, Differential expression of alpha-2,3-sialyltransferases and alpha-1,3/4-fucosyltransferases regulates the levels of sialyl Lewis a and sialyl Lewis x in gastrointestinal carcinoma cells, Int J Biochem Cell Biol, 42 (2010) 80-89.

[194] C. Vazquez-Martin, E. Cuevas, E. Gil-Martin, A. Fernandez-Briera, Correlation analysis between tumor-associated antigen sialyl-Tn expression and ST6GalNAc I activity in human colon adenocarcinoma, Oncology, 67 (2004) 159-165.

[195] S. Pinho, N.T. Marcos, B. Ferreira, A.S. Carvalho, M.J. Oliveira, F. Santos-Silva, A. Harduin-Lepers, C.A. Reis, Biological significance of cancer-associated sialyl-Tn antigen: modulation of malignant phenotype in gastric carcinoma cells, Cancer Lett, 249 (2007) 157-170.

[196] K. Miyazaki, K. Sakuma, Y.I. Kawamura, M. Izawa, K. Ohmori, M. Mitsuki, T. Yamaji, Y. Hashimoto, A. Suzuki, Y. Saito, T. Dohi, R. Kannagi, Colonic epithelial cells express specific ligands for mucosal macrophage immunosuppressive receptors siglec-7 and -9, J Immunol, 188 (2012) 4690-4700.

[197] S. Julien, P.A. Videira, P. Delannoy, Sialyl-tn in cancer: (how) did we miss the target?, Biomolecules, 2 (2012) 435-466.

[198] J. Munkley, The Role of Sialyl-Tn in Cancer, Int J Mol Sci, 17 (2016) 275.

[199] S. Julien, E. Adriaenssens, K. Ottenberg, A. Furlan, G. Courtand, A.S. Vercoutter-Edouart, F.G. Hanisch, P. Delannoy, X. Le Bourhis, ST6GalNAc I expression in MDA-MB-231 breast cancer cells greatly modifies their O-glycosylation pattern and enhances their tumourigenicity, Glycobiology, 16 (2006) 54-64.

[200] S. Julien, M.A. Krzewinski-Recchi, A. Harduin-Lepers, V. Gouyer, G. Huet, X. Le Bourhis, P. Delannoy, Expression of sialyl-Tn antigen in breast cancer cells transfected with the human CMP-Neu5Ac: GalNAc alpha2,6-sialyltransferase (ST6GalNac I) cDNA, Glycoconj J, 18 (2001) 883-893.

[201] M.A. Carrascal, P.F. Severino, M. Guadalupe Cabral, M. Silva, J.A. Ferreira, F. Calais, H. Quinto, C. Pen, D. Ligeiro, L.L. Santos, F. Dall'Olio, P.A. Videira, Sialyl Tn-expressing bladder cancer cells induce a tolerogenic phenotype in innate and adaptive immune cells, Mol Oncol, 8 (2014) 753-765.

[202] E.M. Yazawa, J.E. Geddes-Sweeney, F. Cedeno-Laurent, K.C. Walley, S.R. Barthel, M.J. Opperman, J. Liang, J.Y. Lin, T. Schatton, A.C. Laga, M.C. Mihm, A.A. Qureshi, H.R. Widlund, G.F. Murphy, C.J. Dimitroff, Melanoma Cell Galectin-1 Ligands Functionally Correlate with Malignant Potential, J Invest Dermatol, 135 (2015) 1849-1862.

[203] E. Ong, J. Nakayama, K. Angata, L. Reyes, T. Katsuyama, Y. Arai, M. Fukuda, Developmental regulation of polysialic acid synthesis in mouse directed by two polysialyltransferases, PST and STX, Glycobiology, 8 (1998) 415-424.

[204] M. Suzuki, M. Suzuki, J. Nakayama, A. Suzuki, K. Angata, S. Chen, K. Sakai, K. Hagihara, Y. Yamaguchi, M. Fukuda, Polysialic acid facilitates tumor invasion by glioma cells, Glycobiology, 15 (2005) 887-894.

[205] L. Gong, X. Zhou, J. Yang, Y. Jiang, H. Yang, Effects of the regulation of polysialyltransferase ST8SiaII on the invasiveness and metastasis of small cell lung cancer cells, Oncol Rep, 37 (2017) 131-138.

[206] X. Zhang, W. Dong, H. Zhou, H. Li, N. Wang, X. Miao, L. Jia, alpha-2,8-Sialyltransferase Is Involved in the Development of Multidrug Resistance via PI3K/Akt Pathway in Human Chronic Myeloid Leukemia, IUBMB Life, 67 (2015) 77-87.

[207] K. Koseki, T. Wada, M. Hosono, K. Hata, K. Yamaguchi, K. Nitta, T. Miyagi, Human cytosolic sialidase NEU2-low general tissue expression but involvement in PC-3 prostate cancer cell survival, Biochem Biophys Res Commun, 428 (2012) 142-149.

[208] X. Li, L. Zhang, Y. Shao, Z. Liang, C. Shao, B. Wang, B. Guo, N. Li, X. Zhao, Y. Li, D. Xu, Effects of a human plasma membrane-associated sialidase siRNA on prostate cancer invasion, Biochem Biophys Res Commun, 416 (2011) 270-276.

[209] C. Tringali, I. Silvestri, F. Testa, P. Baldassari, L. Anastasia, R. Mortarini, A. Anichini, A. Lopez-Requena, G. Tettamanti, B. Venerando, Molecular subtyping of metastatic melanoma based on cell ganglioside metabolism profiles, BMC Cancer, 14 (2014) 560.

[210] M. Miyata, M. Kambe, O. Tajima, S. Moriya, H. Sawaki, H. Hotta, Y. Kondo, H. Narimatsu, T. Miyagi, K. Furukawa, K. Furukawa, Membrane sialidase NEU3 is highly expressed in human melanoma cells promoting cell growth with minimal changes in the composition of gangliosides, Cancer Sci, 102 (2011) 2139-2149.

[211] H. Nomura, Y. Tamada, T. Miyagi, A. Suzuki, M. Taira, N. Suzuki, N. Susumu, T. Irimura, D. Aoki, Expression of NEU3 (plasma membrane-associated sialidase) in clear cell adenocarcinoma of the ovary: its relationship with T factor of pTNM classification, Oncol Res, 16 (2006) 289-297.

[212] A. Raz, R. Goldman, I. Yuli, M. Inbar, Isolation of Plasma-Membrane Fragments and Vesicles from Ascites-Fluid of Lymphoma-Bearing Mice and Their Possible Role in Escape Mechanism of Tumors from Host Immune Rejection, Cancer Immunol Immun, 4 (1978) 53-59.

[213] Y. Gu, J. Zhang, W. Mi, J. Yang, F. Han, X. Lu, W. Yu, Silencing of GM3 synthase suppresses lung metastasis of murine breast cancer cells, Breast Cancer Res, 10 (2008) R1.

[214] S. Ladisch, H. Becker, L. Ulsh, Immunosuppression by human gangliosides: I. Relationship of carbohydrate structure to the inhibition of T cell responses, Biochim Biophys Acta, 1125 (1992) 180-188.

[215] R.T. Almaraz, Y. Tian, R. Bhattarcharya, E. Tan, S.H. Chen, M.R. Dallas, L. Chen, Z. Zhang, H. Zhang, K. Konstantopoulos, K.J. Yarema, Metabolic flux increases glycoprotein sialylation: implications for cell adhesion and cancer metastasis, Mol Cell Proteomics, 11 (2012) M112 017558.

[216] K.S. Lau, E.A. Partridge, A. Grigorian, C.I. Silvescu, V.N. Reinhold, M. Demetriou, J.W. Dennis, Complex N-glycan number and degree of branching cooperate to regulate cell proliferation and differentiation, Cell, 129 (2007) 123-134.

[217] H.A. Badr, D.M. AlSadek, M.P. Mathew, C.Z. Li, L.B. Djansugurova, K.J. Yarema, H. Ahmed, Nutrient-deprived cancer cells preferentially use sialic acid to maintain cell surface glycosylation, Biomaterials, 70 (2015) 23-36.

[218] C.T. Saeui, A.V. Nairn, M. Galizzi, C. Douville, P. Gowda, M. Park, V. Dharmarha, S.R. Shah, A. Clarke, M. Austin, K.W. Moremen, K.J. Yarema, Integration of genetic and metabolic features related to sialic acid metabolism distinguishes human breast cell subtypes, PLoS One, 13 (2018) e0195812.

[219] P. Allain, E. Olivier, A. Le Bouil, C. Benoit, P. Geslin, A. Tadei, [Increase of sialic acid concentration in the plasma of patients with coronary disease], Presse Med, 25 (1996) 96-98.

[220] I. Wakabayashi, K. Sakamoto, S. Yoshimoto, E. Kakishita, Serum sialic acid concentration and atherosclerotic risk factors, J Atheroscler Thromb, 1 (1994) 113-117.

[221] M. Haq, S. Haq, P. Tutt, M. Crook, Serum total sialic acid and lipid-associated sialic acid in normal individuals and patients with myocardial infarction, and their relationship to acute phase proteins, Ann Clin Biochem, 30 (Pt 4) (1993) 383-386.

[222] M.A. Crook, A. Treloar, M. Haq, P. Tutt, Serum total sialic acid and acute phase proteins in elderly subjects, Eur J Clin Chem Clin Biochem, 32 (1994) 745-747.

[223] P.K. Nigam, Narain, V.S.,Chandra, N., Puri, V.K., Saran, R.K., Dwivedi, S.K. and Hasan, M. , Serum and platelet sialic acid in acute myocardial infarction, Ind. J. Clin. Biochem., 10 (1995) 106-109.

[224] G.F. Watts, M.A. Crook, S. Haq, S. Mandalia, Serum sialic acid as an indicator of change in coronary artery disease, Metabolism, 44 (1995) 147-148.

[225] E.V. Gracheva, N.N. Samovilova, N.K. Golovanova, O.P. Il'inskaya, E.M. Tararak, P.P. Malyshev, V.V. Kukharchuk, N.V. Prokazova, Sialyltransferase activity of human plasma and aortic intima is enhanced in atherosclerosis, Biochim Biophys Acta, 1586 (2002) 123-128.

[226] S. Kelm, R. Schauer, Sialic acids in molecular and cellular interactions, Int Rev Cytol, 175 (1997) 137-240.

[227] K. Pyorala, M. Laakso, M. Uusitupa, Diabetes and atherosclerosis: an epidemiologic view, Diabetes Metab Rev, 3 (1987) 463-524.

[228] H. Baumann, J. Gauldie, The acute phase response, Immunol Today, 15 (1994) 74-80.

[229] E.B. Wu, P. Lumb, J.B. Chambers, M.A. Crook, Plasma sialic acid and coronary artery atheromatous load in patients with stable chest pain, Atherosclerosis, 145 (1999) 261-266.

[230] G. Lindberg, L. Rastam, P. Nilsson-Ehle, A. Lundblad, J. Ranstam, A.R. Folsom, G.L. Burke, Serum sialic acid and sialoglycoproteins in asymptomatic carotid artery atherosclerosis. ARIC Investigators. Atherosclerosis Risk in Communities, Atherosclerosis, 146 (1999) 65-69.

[231] V.V. Tertov, I.A. Sobenin, Z.A. Gabbasov, E.G. Popov, O. Jaakkola, T. Solakivi, T. Nikkari, V.N. Smirnov, A.N. Orekhov, Multiple-modified desialylated low density lipoproteins that cause intracellular lipid accumulation. Isolation, fractionation and characterization, Lab Invest, 67 (1992) 665-675.

[232] A.N. Orekhov, V.V. Tertov, D.N. Mukhin, I.A. Mikhailenko, Modification of low density lipoprotein by desialylation causes lipid accumulation in cultured cells: discovery of desialylated lipoprotein with altered cellular metabolism in the blood of atherosclerotic patients, Biochem Biophys Res Commun, 162 (1989) 206-211.

[233] A.N. Orekhov, V.V. Tertov, D.N. Mukhin, Desialylated low density lipoprotein-- naturally occurring modified lipoprotein with atherogenic potency, Atherosclerosis, 86 (1991) 153-161.

[234] V.V. Tertov, V.V. Kaplun, I.A. Sobenin, A.N. Orekhov, Low-density lipoprotein modification occurring in human plasma possible mechanism of in vivo lipoprotein desialylation as a primary step of atherogenic modification, Atherosclerosis, 138 (1998) 183-195.

[235] H.S. Kruth, The fate of lipoprotein cholesterol entering the arterial wall, Curr Opin Lipidol, 8 (1997) 246-252.

[236] E.A. Sprague, M. Moser, E.H. Edwards, C.J. Schwartz, Stimulation of receptor-mediated low density lipoprotein endocytosis in neuraminidase-treated cultured bovine aortic endothelial cells, J Cell Physiol, 137 (1988) 251-262.

[237] I. Filipovic, E. Buddecke, Desialized low-density lipoprotein regulates cholesterol metabolism in receptor-deficient fibroblasts, Eur J Biochem, 101 (1979) 119-122.

[238] B. Chappey, B. Beyssen, E. Foos, F. Ledru, J.L. Guermonprez, J.C. Gaux, I. Myara, Sialic acid content of LDL in coronary artery disease: no evidence of desialylation in subjects with coronary stenosis and increased levels in subjects with extensive atherosclerosis and acute myocardial infarction: relation between

desialylation and in vitro peroxidation, Arterioscler Thromb Vasc Biol, 18 (1998) 876-883.

[239] M. Bardor, D.H. Nguyen, S. Diaz, A. Varki, Mechanism of uptake and incorporation of the non-human sialic acid N-glycolylneuraminic acid into human cells, J Biol Chem, 280 (2005) 4228-4237.

[240] A.N. Samraj, O.M. Pearce, H. Laubli, A.N. Crittenden, A.K. Bergfeld, K. Banda, C.J. Gregg, A.E. Bingman, P. Secrest, S.L. Diaz, N.M. Varki, A. Varki, A red meat-derived glycan promotes inflammation and cancer progression, Proc Natl Acad Sci U S A, 112 (2015) 542-547.

[241] B. Byrne, G.G. Donohoe, R. O'Kennedy, Sialic acids: carbohydrate moieties that influence the biological and physical properties of biopharmaceutical proteins and living cells, Drug Discov Today, 12 (2007) 319-326.

[242] Y. Pilatte, J. Bignon, C.R. Lambre, Sialic acids as important molecules in the regulation of the immune system: pathophysiological implications of sialidases in immunity, Glycobiology, 3 (1993) 201-218.

[243] D.T. Fearon, Regulation by membrane sialic acid of beta1H-dependent decay-dissociation of amplification C3 convertase of the alternative complement pathway, Proc Natl Acad Sci U S A, 75 (1978) 1971-1975.

[244] M.K. Pangburn, H.J. Muller-Eberhard, Complement C3 convertase: cell surface restriction of beta1H control and generation of restriction on neuraminidase-treated cells, Proc Natl Acad Sci U S A, 75 (1978) 2416-2420.

[245] S. Meri, M.K. Pangburn, Discrimination between activators and nonactivators of the alternative pathway of complement: regulation via a sialic acid/polyanion binding site on factor H, Proc Natl Acad Sci U S A, 87 (1990) 3982-3986.

[246] S. Ram, A.K. Sharma, S.D. Simpson, S. Gulati, D.P. McQuillen, M.K. Pangburn, P.A. Rice, A novel sialic acid binding site on factor H mediates serum resistance of sialylated Neisseria gonorrhoeae, J Exp Med, 187 (1998) 743-752.

[247] R.P. McEver, Selectins: novel receptors that mediate leukocyte adhesion during inflammation, Thromb Haemost, 65 (1991) 223-228.

[248] R.D. Cummings, D.F. Smith, The selectin family of carbohydrate-binding proteins: structure and importance of carbohydrate ligands for cell adhesion, Bioessays, 14 (1992) 849-856.

[249] S.D. Rosen, Robert Feulgen Lecture 1993. L-selectin and its biological ligands, Histochemistry, 100 (1993) 185-191.

[250] A. Varki, Selectin ligands, Proc Natl Acad Sci U S A, 91 (1994) 7390-7397.

[251] L.A. Lasky, Selectin-carbohydrate interactions and the initiation of the inflammatory response, Annu Rev Biochem, 64 (1995) 113-139.

[252] C. Mandal, C. Mandal, Sialic acid binding lectins, Experientia, 46 (1990) 433-441.

[253] C. Traving, R. Schauer, Structure, function and metabolism of sialic acids, Cell Mol Life Sci, 54 (1998) 1330-1349.

[254] P. Isa, C.F. Arias, S. Lopez, Role of sialic acids in rotavirus infection, Glycoconj J, 23 (2006) 27-37.

[255] G.V. Gee, A.S. Dugan, N. Tsomaia, D.F. Mierke, W.J. Atwood, The role of sialic acid in human polyomavirus infections, Glycoconj J, 23 (2006) 19-26.

[256] F. Lehmann, E. Tiralongo, J. Tiralongo, Sialic acid-specific lectins: occurrence, specificity and function, Cell Mol Life Sci, 63 (2006) 1331-1354.

[257] A.L. Lewis, N. Desa, E.E. Hansen, Y.A. Knirel, J.I. Gordon, P. Gagneux, V. Nizet, A. Varki, Innovations in host and microbial sialic acid biosynthesis revealed by phylogenomic prediction of nonulosonic acid structure, Proc Natl Acad Sci U S A, 106 (2009) 13552-13557.

[258] E. Severi, D.W. Hood, G.H. Thomas, Sialic acid utilization by bacterial pathogens, Microbiology (Reading), 153 (2007) 2817-2822.

[259] E. Vimr, C. Lichtensteiger, To sialylate, or not to sialylate: that is the question, Trends Microbiol, 10 (2002) 254-257.

[260] S. Gordon, The macrophage, Bioessays, 17 (1995) 977-986.

[261] A. Mantovani, A. Sica, S. Sozzani, P. Allavena, A. Vecchi, M. Locati, The chemokine system in diverse forms of macrophage activation and polarization, Trends Immunol, 25 (2004) 677-686.

[262] F.O. Martinez, S. Gordon, The M1 and M2 paradigm of macrophage activation: time for reassessment, F1000Prime Rep, 6 (2014) 13.

[263] C.D. Mills, K. Kincaid, J.M. Alt, M.J. Heilman, A.M. Hill, M-1/M-2 macrophages and the Th1/Th2 paradigm, J Immunol, 164 (2000) 6166-6173.

[264] A. Hartnell, J. Steel, H. Turley, M. Jones, D.G. Jackson, P.R. Crocker, Characterization of human sialoadhesin, a sialic acid binding receptor expressed by resident and inflammatory macrophage populations, Blood, 97 (2001) 288-296.

[265] P.R. Crocker, E.A. Clark, M. Filbin, S. Gordon, Y. Jones, J.H. Kehrl, S. Kelm, N. Le Douarin, L. Powell, J. Roder, R.L. Schnaar, D.C. Sgroi, K. Stamenkovic, R. Schauer, M. Schachner, T.K. van den Berg, P.A. van der Merwe, S.M. Watt, A. Varki, Siglecs: a family of sialic-acid binding lectins, Glycobiology, 8 (1998) v.

[266] S. Kelm, A. Pelz, R. Schauer, M.T. Filbin, S. Tang, M.E. de Bellard, R.L. Schnaar, J.A. Mahoney, A. Hartnell, P. Bradfield, et al., Sialoadhesin, myelin-associated glycoprotein and CD22 define a new family of sialic acid-dependent adhesion molecules of the immunoglobulin superfamily, Curr Biol, 4 (1994) 965-972.

[267] S. Kelm, R. Schauer, J.C. Manuguerra, H.J. Gross, P.R. Crocker, Modifications of cell surface sialic acids modulate cell adhesion mediated by sialoadhesin and CD22, Glycoconj J, 11 (1994) 576-585.

[268] S. Kelm, R. Brossmer, R. Isecke, H.J. Gross, K. Strenge, R. Schauer, Functional groups of sialic acids involved in binding to siglecs (sialoadhesins) deduced from interactions with synthetic analogues, Eur J Biochem, 255 (1998) 663-672.

[269] V.G. Monteiro, C.S. Lobato, A.R. Silva, D.V. Medina, M.A. de Oliveira, S.H. Seabra, W. de Souza, R.A. DaMatta, Increased association of Trypanosoma cruzi with sialoadhesin positive mice macrophages, Parasitol Res, 97 (2005) 380-385.

[270] A.L. Bartlett, T. Grewal, E. De Angelis, S. Myers, K.K. Stanley, Role of the macrophage galactose lectin in the uptake of desialylated LDL, Atherosclerosis, 153 (2000) 219-230.

[271] J.J. Lyons, J.D. Milner, S.D. Rosenzweig, Glycans Instructing Immunity: The Emerging Role of Altered Glycosylation in Clinical Immunology, Front Pediatr, 3 (2015) 54.

[272] A. Varki, P. Gagneux, Multifarious roles of sialic acids in immunity, Ann Ny Acad Sci, 1253 (2012) 16-36.

[273] X.H. Zhang, Q.M. Wang, J.M. Zhang, F.E. Feng, F.R. Wang, H. Chen, Y.Y. Zhang, Y.H. Chen, W. Han, L.P. Xu, K.Y. Liu, X.J. Huang, Desialylation is associated with apoptosis and phagocytosis of platelets in patients with prolonged isolated thrombocytopenia after allo-HSCT, J Hematol Oncol, 8 (2015) 116.

[274] Y. Kaneko, F. Nimmerjahn, J.V. Ravetch, Anti-inflammatory activity of immunoglobulin G resulting from Fc sialylation, Science, 313 (2006) 670-673.

[275] C. Cowing, J.M. Chapdelaine, T cells discriminate between Ia antigens expressed on allogeneic accessory cells and B cells: a potential function for carbohydrate side chains on Ia molecules, Proc Natl Acad Sci U S A, 80 (1983) 6000-6004.

[276] C.J. Boog, J.J. Neefjes, J. Boes, H.L. Ploegh, C.J. Melief, Specific immune responses restored by alteration in carbohydrate chains of surface molecules on antigen-presenting cells, Eur J Immunol, 19 (1989) 537-542.

[277] H.J. Crespo, M.G. Cabral, A.V. Teixeira, J.T. Lau, H. Trindade, P.A. Videira, Effect of sialic acid loss on dendritic cell maturation, Immunology, 128 (2009) e621-631.

[278] Y. Watanabe, A. Shiratsuchi, K. Shimizu, T. Takizawa, Y. Nakanishi, Role of phosphatidylserine exposure and sugar chain desialylation at the surface of influenza virus-infected cells in efficient phagocytosis by macrophages, J Biol Chem, 277 (2002) 18222-18228.

[279] Y. Watanabe, A. Shiratsuchi, K. Shimizu, T. Takizawa, Y. Nakanishi, Stimulation of phagocytosis of influenza virus-infected cells through surface desialylation of macrophages by viral neuraminidase, Microbiol Immunol, 48 (2004) 875-881.

[280] D.H. Allendorf, E.H. Franssen, G.C. Brown, Lipopolysaccharide activates microglia via neuraminidase 1 desialylation of Toll-like Receptor 4, J Neurochem, 155 (2020) 403-416.

[281] E.L. Hess, A.F. Coburn, R.C. Bates, P. Murphy, A new method for measuring sialic acid levels in serum and its application to rheumatic fever, J Clin Invest, 36 (1957) 449-455.

[282] S. Rajaram, B.P. Danasekaran, R. Venkatachalapathy, K.V. Prashad, S. Rajaram, N-acetylneuraminic acid: A scrutinizing tool in oral squamous cell carcinoma diagnosis, Dent Res J (Isfahan), 14 (2017) 267-271.

[283] S. Habibi, H. Jamshidian, M. Kadivar, M.R. Eshraghian, M.H. Javanbakht, H. Derakhshanian, M. Djalali, A study of lipid- and protein- bound sialic acids for the diagnosis of bladder cancer and their relationships with the severity of malignancy, Rep Biochem Mol Biol, 2 (2014) 70-75.

[284] S.M. Goodarzi MT, Nomani H, Shahriarahmadi A., Relationship between total and lipid-bound serum sialic acid and some tumor markers, Iran J Med Sci. , 30 (2015) 124-127.

[285] B.S. Krishnan K, Estimation of total and lipid bound sialic acid in serum in oral leukoplakia. , J Clin Diagnostic Res., 11 (2017) ZC25–ZC27.

[286] J. Cheeseman, G. Kuhnle, D.I.R. Spencer, H.M.I. Osborn, Assays for the identification and quantification of sialic acids: Challenges, opportunities and future perspectives, Bioorg Med Chem, 30 (2021) 115882.

[287] W. Ayala, L.V. Moore, E.L. Hess, The purple color reaction given by diphenylamine reagent. I. With normal and rheumatic fever sera, J Clin Invest, 30 (1951) 781-785.

[288] I. Werner, L. Odin, On the presence of sialic acid in certain glycoproteins and in gangliosides, Acta Soc Med Ups, 57 (1952) 230-241.

[289] H. Hess, E. Rolde, Fluorometric Assay of Sialic Acid in Brain Gangliosides, J Biol Chem, 239 (1964) 3215-3220.

[290] J. Folch, S. Arsove, J.A. Meath, Isolation of brain strandin, a new type of large molecule tissue component, J Biol Chem, 191 (1951) 819-831.

[291] L.H. Klenk E, Orcinol method for measuring sialic acid, Hoppe-Seyler's Z Physiol Chem, 270 (1941) 185-193.

[292] B.P. Peters, N.N. Aronson, Jr., Reactivity of the sialic acid derivative 5-acetamido-3, 5-dideoxy-L-arabino-heptulosonic acid in the resorcinol and thiobarbituric acid assays, Carbohydr Res, 47 (1976) 345-353.

[293] R. Schauer, J.P. Kamerling, Exploration of the Sialic Acid World, Adv Carbohydr Chem Biochem, 75 (2018) 1-213.

[294] L. Svennerholm, Quantitative estimation of sialic acids. II. A colorimetric resorcinol-hydrochloric acid method, Biochim Biophys Acta, 24 (1957) 604-611.

[295] L. Svennerholm, Assay of sialic acids, Methods in Enzymology, 6 (1963) 459-462.

[296] D. Aminoff, The determination of free sialic acid in the presence of the bound compound, Virology, 7 (1959) 355-357.

[297] L. Warren, The thiobarbituric acid assay of sialic acids, J Biol Chem, 234 (1959) 1971-1975.

[298] Y. Massamiri, G. Durand, A. Richard, J. Feger, J. Agneray, Determination of erythrocyte surface sialic acid residues by a new colorimetric method, Anal Biochem, 97 (1979) 346-351.

[299] K.S. Hammond, D.S. Papermaster, Fluorometric assay of sialic acid in the picomole range: a modification of the thiobarbituric acid assay, Anal Biochem, 74 (1976) 292-297.

[300] J.I. Murayama, M. Tomita, A. Tsuji, A. Hamada, Fluorimetric assay of sialic acids, Anal Biochem, 73 (1976) 535-538.

[301] A.K. Shukla, R. Schauer, Fluorimetric determination of unsubstituted and 9(8)-O-acetylated sialic acids in erythrocyte membranes, Hoppe Seylers Z Physiol Chem, 363 (1982) 255-262.

[302] K. Matsuno, S. Suzuki, Simple fluorimetric method for quantification of sialic acids in glycoproteins, Anal Biochem, 375 (2008) 53-59.

[303] P. Brunetti, G.W. Jourdian, S. Roseman, The sialic acids. III. Distribution and properties of animal N-acetylneuraminic aldolase, J Biol Chem, 237 (1962) 2447-2453.

[304] K. Sugahara, K. Sugimoto, O. Nomura, T. Usui, Enzymatic assay of serum sialic acid, Clin Chim Acta, 108 (1980) 493-498.

[305] S. Teshima, K. Tamai, Y. Hayashi, S. Emi, New enzymatic determination of sialic acid in serum, Clin Chem, 34 (1988) 2291-2294.

[306] M.A. Maliakal, M.H. Ravindranath, R.F. Irie, D.L. Morton, An improved method for the measurement of total lipid-bound sialic acids after cleavage of alpha 2,8 sialic acid linkage with Vibrio cholerae sialidase in the presence of cholic acid, SDS and Ca2+, Glycoconj J, 11 (1994) 97-104.

[307] B. Yesilyurt, U. Sahar, R. Deveci, Determination of the type and quantity of sialic acid in the egg jelly coat of the sea urchin Paracentrotus lividus using capillary LC-ESI-MS/MS, Mol Reprod Dev, 82 (2015) 115-122.

[308] S. Cavdarli, J.H. Dewald, N. Yamakawa, Y. Guerardel, M. Terme, J.M. Le Doussal, P. Delannoy, S. Groux-Degroote, Identification of 9-O-acetyl-N-acetylneuraminic acid (Neu5,9Ac2) as main O-acetylated sialic acid species of GD2 in breast cancer cells, Glycoconj J, 36 (2019) 79-90.

[309] R.I. Thomson, R.A. Gardner, K. Strohfeldt, D.L. Fernandes, G.P. Stafford, D.I.R. Spencer, H.M.I. Osborn, Analysis of Three Epoetin Alpha Products by LC and LC-MS Indicates Differences in Glycosylation Critical Quality Attributes, Including Sialic Acid Content, Anal Chem, 89 (2017) 6455-6462.

[310] D.C. Hurum, J.S. Rohrer, Determination of sialic acids in infant formula by chromatographic methods: a comparison of high-performance anion-exchange chromatography with pulsed amperometric detection and ultra-high-performance liquid chromatography methods, J Dairy Sci, 95 (2012) 1152-1161.

[311] K.T. Tang, L.N. Liang, Y.Q. Cai, S.F. Mou, Determination of Sialic Acid in Milk and Products Using High Performance Anion-Exchange Chromatography

Coupled with Pulsed Amperometric Detection, Chinese J Anal Chem, 36 (2008) 1535-1538.

[312] A.K. Shukla, N. Scholz, E.H. Reimerdes, R. Schauer, High-performance liquid chromatography of N,O-acylated sialic acids, Anal Biochem, 123 (1982) 78-82.

[313] A.K. Shukla, R. Schauer, Analysis of N,O-Acylated Neuraminic Acids by High-Performance Liquid Anion-Exchange Chromatography, Journal of Chromatography, 244 (1982) 81-89.

[314] Y.A. Kazunori KOBAYASHI, Kayo KAWAGUCHI, Shinzo TANABE, Toshio IMANARI, Fluorometric Determination of N-Acetyl and N-Glycolyl Neuraminic Acids by High Performance Liquid Chromatography as Their 4'-Hydrazino-2-stilbazole Derivatives, Analytical Sciences, 1 (1985) 81-84.

[315] S. Honda, S. Iwase, S. Suzuki, K. Kakehi, Fluorometric determination of sialic acids using malononitrile in weakly alkaline media and its application to postcolumn labeling in high-performance liquid chromatography, Anal Biochem, 160 (1987) 455-461.

[316] K. Li, Determination of sialic acids in human serum by reversed-phase liquid chromatography with fluorimetric detection, J Chromatogr, 579 (1992) 209-213.

[317] K.R. Anumula, Rapid quantitative determination of sialic acids in glycoproteins by high-performance liquid chromatography with a sensitive fluorescence detection, Anal Biochem, 230 (1995) 24-30.

[318] P.G. Stanton, Z. Shen, E.A. Kecorius, P.G. Burgon, D.M. Robertson, M.T. Hearn, Application of a sensitive HPLC-based fluorometric assay to determine the sialic

acid content of human gonadotropin isoforms, J Biochem Biophys Methods, 30 (1995) 37-48.

[319] S. Hara, M. Yamaguchi, Y. Takemori, M. Nakamura, Y. Ohkura, Highly sensitive determination of N-acetyl- and N-glycolylneuraminic acids in human serum and urine and rat serum by reversed-phase liquid chromatography with fluorescence detection, J Chromatogr, 377 (1986) 111-119.

[320] M.J. Martin, E. Vazquez, R. Rueda, Application of a sensitive fluorometric HPLC assay to determine the sialic acid content of infant formulas, Anal Bioanal Chem, 387 (2007) 2943-2949.

[321] V. Spichtig, J. Michaud, S. Austin, Determination of sialic acids in milks and milk-based products, Anal Biochem, 405 (2010) 28-40.

[322] S. Hara, M. Yamaguchi, Y. Takemori, K. Furuhata, H. Ogura, M. Nakamura, Determination of mono-O-acetylated N-acetylneuraminic acids in human and rat sera by fluorometric high-performance liquid chromatography, Anal Biochem, 179 (1989) 162-166.

[323] L.B. Wang, D. Wang, X. Zhou, L.J. Wu, X.L. Sun, Systematic investigation of quinoxaline derivatization of sialic acids and their quantitation applicability using high performance liquid chromatography, Rsc Adv, 4 (2014) 45797-45803.

[324] D. Wang, X. Zhou, L. Wang, S. Wang, X.L. Sun, Quantification of free sialic acid in human plasma through a robust quinoxalinone derivatization and LC-MS/MS using isotope-labeled standard calibration, J Chromatogr B Analyt Technol Biomed Life Sci, 944 (2014) 75-81.

[325] C.J. Shaw, H. Chao, B. Xiao, Determination of sialic acids by liquid chromatography-mass spectrometry, J Chromatogr A, 913 (2001) 365-370.

[326] Y. Zhao, G. Mahajan, C.R. Kothapalli, X.L. Sun, Sialylation status and mechanical properties of THP-1 macrophages upon LPS stimulation, Biochem Biophys Res Commun, 518 (2019) 573-578.

[327] M. van der Ham, B.H. Prinsen, J.G. Huijmans, N.G. Abeling, B. Dorland, R. Berger, T.J. de Koning, M.G. de Sain-van der Velden, Quantification of free and total sialic acid excretion by LC-MS/MS, J Chromatogr B Analyt Technol Biomed Life Sci, 848 (2007) 251-257.

[328] S.P. Galuska, H. Geyer, C. Bleckmann, R.C. Rohrich, K. Maass, A.K. Bergfeld, M. Muhlenhoff, R. Geyer, Mass spectrometric fragmentation analysis of oligosialic and polysialic acids, Anal Chem, 82 (2010) 2059-2066.

[329] G. Palmisano, M.R. Larsen, N.H. Packer, M. Thaysen-Andersen, Structural analysis of glycoprotein sialylation -part II: LC-MS based detection, Rsc Adv, 3 (2013) 22706-22726.

[330] K.R. Reiding, A. Bondt, R. Hennig, R.A. Gardner, R. O'Flaherty, I. Trbojevic-Akmacic, A. Shubhakar, J.M.W. Hazes, U. Reichl, D.L. Fernandes, M. Pucic-Bakovic, E. Rapp, D.I.R. Spencer, R. Dolhain, P.M. Rudd, G. Lauc, M. Wuhrer, High-throughput Serum N-Glycomics: Method Comparison and Application to Study Rheumatoid Arthritis and Pregnancy-associated Changes, Mol Cell Proteomics, 18 (2019) 3-15.

[331] T. Nishikaze, Sialic acid derivatization for glycan analysis by mass spectrometry, Proc Jpn Acad Ser B Phys Biol Sci, 95 (2019) 523-537.

[332] T. Nishikaze, H. Tsumoto, S. Sekiya, S. Iwamoto, Y. Miura, K. Tanaka, Differentiation of Sialyl Linkage Isomers by One-Pot Sialic Acid Derivatization for Mass Spectrometry-Based Glycan Profiling, Anal Chem, 89 (2017) 2353-2360.

[333] S. Holst, B. Heijs, N. de Haan, R.J. van Zeijl, I.H. Briaire-de Bruijn, G.W. van Pelt, A.S. Mehta, P.M. Angel, W.E. Mesker, R.A. Tollenaar, R.R. Drake, J.V. Bovee, L.A. McDonnell, M. Wuhrer, Linkage-Specific in Situ Sialic Acid Derivatization for N-Glycan Mass Spectrometry Imaging of Formalin-Fixed Paraffin-Embedded Tissues, Anal Chem, 88 (2016) 5904-5913.

[334] N. Suzuki, T. Abe, S. Natsuka, Quantitative LC-MS and MS/MS analysis of sialylated glycans modified by linkage-specific alkylamidation, Anal Biochem, 567 (2019) 117-127.

[335] M. Guha, N. Mackman, LPS induction of gene expression in human monocytes, Cell Signal, 13 (2001) 85-94.

[336] S. Chiu, A. Bharat, Role of monocytes and macrophages in regulating immune response following lung transplantation, Curr Opin Organ Transplant, 21 (2016) 239-245.

[337] S. Tsuchiya, M. Yamabe, Y. Yamaguchi, Y. Kobayashi, T. Konno, K. Tada, Establishment and characterization of a human acute monocytic leukemia cell line (THP-1), Int J Cancer, 26 (1980) 171-176.

[338] S. Gordon, P.R. Taylor, Monocyte and macrophage heterogeneity, Nat Rev Immunol, 5 (2005) 953-964.

[339] M. Desjardins, L.A. Huber, R.G. Parton, G. Griffiths, Biogenesis of phagolysosomes proceeds through a sequential series of interactions with the endocytic apparatus, J Cell Biol, 124 (1994) 677-688.

[340] M. Daigneault, J.A. Preston, H.M. Marriott, M.K. Whyte, D.H. Dockrell, The identification of markers of macrophage differentiation in PMA-stimulated THP-1 cells and monocyte-derived macrophages, PLoS One, 5 (2010) e8668.

[341] H.B. Fleit, C.D. Kobasiuk, The human monocyte-like cell line THP-1 expresses Fc gamma RI and Fc gamma RII, J Leukoc Biol, 49 (1991) 556-565.

[342] J. Kopitz, C. Muhl, V. Ehemann, C. Lehmann, M. Cantz, Effects of cell surface ganglioside sialidase inhibition on growth control and differentiation of human neuroblastoma cells, Eur J Cell Biol, 73 (1997) 1-9.

[343] G. Wu, R.W. Ledeen, Stimulation of neurite outgrowth in neuroblastoma cells by neuraminidase: putative role of GM1 ganglioside in differentiation, J Neurochem, 56 (1991) 95-104.

[344] P. Altevogt, M. Fogel, R. Cheingsong-Popov, J. Dennis, P. Robinson, V. Schirrmacher, Different patterns of lectin binding and cell surface sialylation detected on related high- and low-metastatic tumor lines, Cancer Res, 43 (1983) 5138-5144.

[345] C.L. Schengrund, R.N. Lausch, A. Rosenberg, Sialidase activity in transformed cells, J Biol Chem, 248 (1973) 4424-4428.

[346] N.M. Stamatos, S. Curreli, D. Zella, A.S. Cross, Desialylation of glycoconjugates on the surface of monocytes activates the extracellular signal-related kinases ERK

1/2 and results in enhanced production of specific cytokines, J Leukoc Biol, 75 (2004) 307-313.

[347] R.D. Sanderson, S.K. Bandari, I. Vlodavsky, Proteases and glycosidases on the surface of exosomes: Newly discovered mechanisms for extracellular remodeling, Matrix Biol, 75-76 (2019) 160-169.

[348] L. Bezrukov, P.S. Blank, I.V. Polozov, J. Zimmerberg, An adhesion-based method for plasma membrane isolation: evaluating cholesterol extraction from cells and their membranes, Anal Biochem, 394 (2009) 171-176.

[349] C. Oetke, R. Brossmer, L.R. Mantey, S. Hinderlich, R. Isecke, W. Reutter, O.T. Keppler, M. Pawlita, Versatile biosynthetic engineering of sialic acid in living cells using synthetic sialic acid analogues, J Biol Chem, 277 (2002) 6688-6695.

[350] S. Abdulkhalek, S.R. Amith, S.L. Franchuk, P. Jayanth, M. Guo, T. Finlay, A. Gilmour, C. Guzzo, K. Gee, R. Beyaert, M.R. Szewczuk, Neu1 sialidase and matrix metalloproteinase-9 cross-talk is essential for Toll-like receptor activation and cellular signaling, J Biol Chem, 286 (2011) 36532-36549.

[351] J. da Silva Correia, R.J. Ulevitch, MD-2 and TLR4 N-linked glycosylations are important for a functional lipopolysaccharide receptor, J Biol Chem, 277 (2002) 1845-1854.

[352] M.J. Gramer, C.F. Goochee, Glycosidase activities in Chinese hamster ovary cell lysate and cell culture supernatant, Biotechnol Prog, 9 (1993) 366-373.

[353] S. Katoh, T. Miyagi, H. Taniguchi, Y. Matsubara, J. Kadota, A. Tominaga, P.W. Kincade, S. Matsukura, S. Kohno, Cutting edge: an inducible sialidase regulates

the hyaluronic acid binding ability of CD44-bearing human monocytes, J Immunol, 162 (1999) 5058-5061.

[354] K. Ohtsubo, J.D. Marth, Glycosylation in cellular mechanisms of health and disease, Cell, 126 (2006) 855-867.

[355] D.M. Mosser, J.P. Edwards, Exploring the full spectrum of macrophage activation, Nature Reviews Immunology, 8 (2008) 958-969.

[356] F.O. Martinez, L. Helming, S. Gordon, Alternative Activation of Macrophages: An Immunologic Functional Perspective, Annu Rev Immunol, 27 (2009) 451-483.

[357] S.K. Biswas, A. Mantovani, Macrophage plasticity and interaction with lymphocyte subsets: cancer as a paradigm, Nature Immunology, 11 (2010) 889-896.

[358] S.C. Erzurum, G.P. Downey, D.E. Doherty, B. Schwab, E.L. Elson, G.S. Worthen, Mechanisms of Lipopolysaccharide-Induced Neutrophil Retention - Relative Contributions of Adhesive and Cellular Mechanical-Properties, Journal of Immunology, 149 (1992) 154-162.

[359] N.R. Patel, M. Bole, C. Chen, C.C. Hardin, A.T. Kho, J. Mih, L.H. Deng, J. Butler, D. Tschumperlin, J.J. Fredberg, R. Krishnan, H. Koziel, Cell Elasticity Determines Macrophage Function, Plos One, 7 (2012).

[360] I. Kang, Q. Wang, S.J. Eppell, R.E. Marchant, C.M. Doerschuk, Effect of Neutrophil Adhesion on the Mechanical Properties of Lung Microvascular Endothelial Cells, Am J Resp Cell Mol, 43 (2010) 591-598.

[361] L. Vonna, A. Wiedemann, M. Aepfelbacher, E. Sackmann, Micromechanics of filopodia mediated capture of pathogens by macrophages, Eur Biophys J Biophy, 36 (2007) 145-151.

[362] H. Kress, E.H.K. Stelzer, D. Holzer, F. Buss, G. Griffiths, A. Rohrbach, Filopodia act as phagocytic tentacles and pull with discrete steps and a load-dependent velocity, P Natl Acad Sci USA, 104 (2007) 11633-11638.

[363] R.S. Flannagan, R.E. Harrison, C.M. Yip, K. Jaqaman, S. Grinstein, Dynamic macrophage "probing" is required for the efficient capture of phagocytic targets, Journal of Cell Biology, 191 (2010) 1205-1218.

[364] L. Kong, L. Sun, H.X. Zhang, Q. Liu, Y. Liu, L.H. Qin, G.J. Shi, J.H. Hu, A.J. Xu, Y.P. Sun, D.S. Li, Y.F. Shi, J.W. Zang, J. Zhu, Z. Chen, Z.G. Wang, B.X. Ge, An Essential Role for RIG-I in Toll-like Receptor-Stimulated Phagocytosis, Cell Host Microbe, 6 (2009) 150-161.

[365] C.A. Grant, P.C. Twigg, R.F. Saeed, G. Lawson, R.A. Falconer, S.D. Shnyder, The Effect of Polysialic Acid Expression on Glioma Cell Nano-mechanics, Bionanoscience, 6 (2016) 81-84.

[366] I. Sokolov, M.E. Dokukin, N.V. Guz, Method for quantitative measurements of the elastic modulus of biological cells in AFM indentation experiments, Methods, 60 (2013) 202-213.

[367] B. Knoops, S. Becker, M.A. Poncin, J. Glibert, S. Derclaye, A. Clippe, D. Alsteens, Specific Interactions Measured by AFM on Living Cells between Peroxiredoxin-5 and TLR4: Relevance for Mechanisms of Innate Immunity, Cell Chem Biol, 25 (2018) 550-+.

[368] A.V. Pshezhetsky, L.I. Ashmarina, Desialylation of surface receptors as a new dimension in cell signaling, Biochemistry-Moscow+, 78 (2013) 736-745.

[369] W. Chanput, J.J. Mes, H.J. Wichers, THP-1 cell line: An in vitro cell model for immune modulation approach, International Immunopharmacology, 23 (2014) 37-45.

[370] R.M. Hochmuth, J.Y. Shao, J.W. Dai, M.P. Sheetz, Deformation and flow of membrane into tethers extracted from neuronal growth cones, Biophysical Journal, 70 (1996) 358-369.

[371] A. Diz-Munoz, M. Krieg, M. Bergert, I. Ibarlucea-Benitez, D.J. Muller, E. Paluch, C.P. Heisenberg, Control of Directed Cell Migration In Vivo by Membrane-to-Cortex Attachment, Plos Biol, 8 (2010).

[372] A. Varki, N-glycolylneuraminic acid deficiency in humans, Biochimie, 83 (2001) 615-622.

[373] J. Du, M.A. Meledeo, Z.Y. Wang, H.S. Khanna, V.D.P. Paruchuri, K.J. Yarema, Metabolic glycoengineering: Sialic acid and beyond, Glycobiology, 19 (2009) 1382-1401.

[374] C.G. Feng, L. Zhang, C. Nguyen, S.N. Vogel, S.E. Goldblum, W.C. Blackwelder, A.S. Cross, Neuraminidase Reprograms Lung Tissue and Potentiates Lipopolysaccharide-Induced Acute Lung Injury in Mice, Journal of Immunology, 191 (2013) 4828-4837.

[375] J. Pi, T. Li, J.X. Liu, X.H. Su, R. Wang, F. Yang, H.H. Bai, H. Jin, J.Y. Cai, Detection of lipopolysaccharide induced inflammatory responses in RAW264.7 macrophages using atomic force microscope, Micron, 65 (2014) 1-9.

[376] S. Leporatti, A. Gerth, G. Kohler, B. Kohlstrunk, S. Hauschildt, E. Donath, Elasticity and adhesion of resting and lipopolysaccharide-stimulated macrophages, Febs Letters, 580 (2006) 450-454.

[377] A. Aderem, D.M. Underhill, Mechanisms of phagocytosis in macrophages, Annu Rev Immunol, 17 (1999) 593-623.

[378] M. Cohen, A. Varki, The Sialome-Far More Than the Sum of Its Parts, Omics-a Journal of Integrative Biology, 14 (2010) 455-464.

[379] A.K. Bergfeld, O.M.T. Pearce, S.L. Diaz, T. Pham, A. Varki, Metabolism of Vertebrate Amino Sugars with N-Glycolyl Groups ELUCIDATING THE INTRACELLULAR FATE OF THE NON-HUMAN SIALIC ACID N-GLYCOLYLNEURAMINIC ACID, Journal of Biological Chemistry, 287 (2012) 28865-28881.

[380] C. Agatemor, M.J. Buettner, R. Ariss, K. Muthiah, C.T. Saeui, K.J. Yarema, Exploiting metabolic glycoengineering to advance healthcare, Nat Rev Chem, 3 (2019) 605-620.

[381] B. Cheng, R. Xie, L. Dong, X. Chen, Metabolic Remodeling of Cell-Surface Sialic Acids: Principles, Applications, and Recent Advances, Chembiochem, 17 (2016) 11-27.

[382] W.M. Kalka-Moll, A.O. Tzianabos, P.W. Bryant, M. Niemeyer, H.L. Ploegh, D.L. Kasper, Zwitterionic polysaccharides stimulate T cells by MHC class II-dependent interactions, Journal of Immunology, 169 (2002) 6149-6153.

[383] R. Isecke, R. Brossmer, Synthesis of 5-N-Thioacylated and 9-N-Thioacylated Sialic Acids, Tetrahedron, 50 (1994) 7445-7460.

[384] T.J. Monica, D.C. Andersen, C.F. Goochee, A mathematical model of sialylation of

N-linked oligosaccharides in the trans-Golgi network, Glycobiology, 7 (1997)

515-521.

9 783384 262264